CAMBRIDGE MONOGRAPHS ON PHYSICS

GENERAL EDITORS

N. FEATHER, F.R.S.

Professor of Natural Philosophy in the University of Edinburgh

D. SHOENBERG, F.R.S.

Fellow of Gonville and Caius College, Cambridge

FERROMAGNETIC DOMAINS

FERROMAGNETIC DOMAINS

BY

K. H. STEWART

M.A., Ph.D.

*Sometime Rouse Ball Student of
Trinity College, Cambridge*

CAMBRIDGE

AT THE UNIVERSITY PRESS

1954

CAMBRIDGE
UNIVERSITY PRESS

University Printing House, Cambridge CB2 8BS, United Kingdom

Cambridge University Press is part of the University of Cambridge.

It furthers the University's mission by disseminating knowledge in the pursuit of education, learning and research at the highest international levels of excellence.

www.cambridge.org
Information on this title: www.cambridge.org/9781107662995

First published 1954
First paperback edition 2016

A catalogue record for this publication is available from the British Library

ISBN 978-1-107-66299-5 Paperback

GENERAL PREFACE

The Cambridge Physical Tracts, out of which this series of Monographs has developed, were planned and originally published in a period when book production was a fairly rapid process. Unfortunately, that is no longer so, and to meet the new situation a change of title and a slight change of emphasis have been decided on. The major aim of the series will still be the presentation of the results of recent research, but individual volumes will be somewhat more substantial, and more comprehensive in scope, than were the volumes of the older series. This will be true, in many cases, of new editions of the Tracts, as these are republished in the expanded series, and it will be true in most cases of the Monographs which have been written since the War or are still to be written.

The aim will be that the series as a whole shall remain representative of the entire field of pure physics, but it will occasion no surprise if, during the next few years, the subject of nuclear physics claims a large share of attention. Only in this way can justice be done to the enormous advances in this field of research over the War years.

N. F.
D. S.

CONTENTS

LIST OF PLATES   *page* ix

PREFACE   xi

CHAPTER I

Introduction

1.1. The domain hypothesis, p. 1. **1.2.** The spontaneous magnetization, p. 3.
1.3. The directions of domain magnetizations, p. 5. **1.4.** Magnetic field,
p. 7. **1.5.** Magnetocrystalline anisotropy, p. 8. **1.6.** Stress anisotropy, p. 9.
1.7. Reasons for the occurrence of domains, p. 10. **1.8.** Changes in domain
arrangements, p. 13. **1.9.** Synopsis of later chapters, p. 15.

CHAPTER II

Magnetocrystalline anisotropy

2.1. Single crystals, p. 17. **2.2.** Magnetization curves in directions of symmetry,
p. 18. **2.3.** Physical origin of the anisotropy, p. 24. **2.4.** Arbitrary orientation.
Normal component of magnetization, p. 25. **2.5.** Polycrystalline specimens,
p. 32. **2.6.** The approach to saturation, p. 34. **2.7.** Measurement of the anisotropy constants, p. 36. **2.8.** Values of the anisotropy constants, p. 40.

CHAPTER III

Magnetostriction

3.1. Introduction, p. 41. **3.2.** Distortion of a cubic crystal by magnetization,
field and stress, p. 42. **3.3.** Volume and linear magnetostriction, p. 44.
3.4. Magnetostrictive energy, p. 47. **3.5.** Variation of magnetostriction with
magnetization, p. 48. **3.6.** Effect of stresses on magnetic properties, p. 53.
3.7. The 'ΔE effect', p. 64. **3.8.** Physical interpretation of magnetostriction, p. 66.

CHAPTER IV

Domain arrangement

4.1. Introduction, p. 67. **4.2.** Orientation of domain walls, p. 67. **4.3.** Closing
domains, p. 70. **4.4.** Survey of factors determining domain arrangement, p. 73.
4.5. Domain arrangements in a single-crystal rod, p. 74. **4.6.** Experimental study
of domain arrangements: 'Bitter patterns', p. 81. **4.7.** Comparison with theory,
p. 87. **4.8.** Mechanism of formation of powder patterns, p. 89. **4.9.** Powder
patterns on other materials, p. 91. **4.10.** Other methods of studying domain
structures, p. 92.

CHAPTER V

Domain walls

5.1. Approximate treatment of wall width and energy, p. 93. **5.2.** Change of spin direction within a wall, p. 94. **5.3.** Energy in walls, p. 95. **5.4.** The width of walls, p. 100. **5.5.** Walls in materials under stress, p. 101. **5.6.** Numerical estimates of wall energy and thickness, p. 102. **5.7.** Experimental measurement of wall energy, p. 103.

CHAPTER VI

Hindrances to domain wall movements

6.1. Introduction, p. 108. **6.2.** The effect of internal stresses on domain volume energy, p. 109. **6.3.** The effect of internal stresses on domain wall energy, p. 110. **6.4.** Effects of 'inclusions', p. 112. **6.5.** Effects of larger amounts of non-ferromagnetics, p. 117. **6.6.** Other types of inclusions: stress centres, p. 120. **6.7.** Applications to real materials, p. 121. **6.8.** The Becker-Kersten stress theory, p. 121. **6.9.** Experiments on the stress theory, p. 127. **6.10.** Imperfections of the stress theory, p. 130. **6.11.** Néel's theory, p. 132. **6.12.** Modification of the stress theory, p. 133. **6.13.** Directional dependence of initial susceptibility, p. 134. **6.14.** Formal treatments of the Rayleigh region, p. 137. **6.15** Practical applications, p. 140.

CHAPTER VII

Time effects

7.1. Introduction, p. 144. **7.2.** Effects of eddy currents, p. 147. **7.3.** Mechanisms producing magnetic viscosity, p. 150. **7.4.** Experiments with simple domain arrangements, p. 157.

CHAPTER VIII

Magnetic and thermal energy changes

8.1. Introduction, p. 161. **8.2.** Irreversible heating effects: hysteresis, p. 162. **8.3.** Mechanisms for disposal of hysteresis energy, p. 164. **8.4.** Reversibility of wall movements, p. 165. **8.5.** Reversible temperature changes, p. 168.

REFERENCES   *page* 171

INDEX   175

LIST OF PLATES

Plates I, II, III are placed between pp. 84–5
Plates IV and V   *between pp.* 88–9
Plates VI and VII   *between pp.* 92–3

All the plates show patterns obtained by applying colloidal magnetite to carefully polished crystal surfaces. Plates I–V show patterns obtained on crystals of 3 % silicon-iron alloy. The directions of the [100] axes of the crystals are indicated below each plate and the scale is shown by arrows representing 1/10 mm.

PLATE I   Patterns on a crystal mechanically polished and electro-lytically polished

II   Pattern on a (100) surface

III   Patterns showing effect of deviation from a (100) plane

IV   Effect of change of magnetization on patterns on a (100) surface

V   Pattern near the corner of a crystal, showing effect of crystal size on domain spacing

VI   Pattern on a (110) surface of a cobalt-nickel crystal

VII   Patterns on a cobalt crystal

Plates III, IV and V appeared in a paper by Williams, Bozorth and Shockley (*Physical Review*, **75**, 155, 1949) and Plates I, II, VI and VII in one by Bozorth (*Journal de Physique et le Radium*, **12**, 308, 1951); they are reproduced here by courtesy of the authors and editors.

PREFACE

If this book had been written ten years ago, it could have been little more than a summary of Becker and Döring's *Ferromagnetismus*. Now, though any account of ferromagnetism must start from the basic ideas they set out so clearly, it is possible to describe much recent work which extends and modifies earlier ideas on ferromagnetic domains. Much that was previously a matter of conjecture and qualitative argument has become, in the last ten years, certain or open to exact discussion, but greater precision has inevitably brought fresh problems to light. This book attempts to give a coherent outline of the fundamentals of domain behaviour but not to give detailed consideration to all parts of the subject. I have naturally selected for fuller treatment those topics in which I have been most directly interested. Many of the subjects dealt with sketchily in the present book are given greater emphasis in K. Hoselitz's *Ferromagnetic Properties of Metals and Alloys*, while for an encyclopaedic account of present knowledge reference can be made to R. M. Bozorth's *Ferromagnetism*; both these books appeared while the present one was in preparation.

I acknowledge gratefully the help I have had from many in writing this book, but above all that from Dr D. Shoenberg, F.R.S., who introduced me to the subject and has given much valuable encouragement and advice at all stages. I am also very grateful to Dr R. M. Bozorth and his colleagues for their generosity in allowing me to use as illustrations some of their beautiful photographs of domain patterns.

K.H.S.

October 1953

CHAPTER I

INTRODUCTION

1.1. The domain hypothesis

The outstanding property of ferromagnetic materials is the ease with which their intensity of magnetization can be varied. There is abundant evidence that the source of their magnetism is the same as in paramagnetic substances, namely, the magnetic moment possessed by certain electrons of the material by virtue of their spin or orbital momentum. In paramagnetics, however, the directions of these moments are arranged at random, so that no magnetization can be detected on a macroscopic scale. In order to produce a measurable magnetization the elementary moments have to be aligned more or less in the same direction, despite the effects of thermal agitation. Theory and experiment agree in showing that very large magnetic fields ($\sim 10^6$ oersted) are needed if the state of perfect alignment, with the macroscopic magnetic moment equal to the sum of the elementary moments, is to be approached at room temperatures.

In ferromagnetics, on the other hand, a marked degree of alignment is somehow achieved in very much smaller fields (~ 100 oersted) and may even persist in the absence of any field at all. The fundamental problem of ferromagnetism is to explain why the elementary moments of a ferromagnetic can be aligned so much more easily than those of a paramagnetic. The first satisfactory explanation, which has been the basis of all later theories, was put forward by Weiss in 1907. He suggested that there were forces of interaction between the elementary magnetic moments, tending to make each one parallel to its neighbours. It is clear that such forces would cause all the moments to be aligned in the same direction at the absolute zero of temperature, and it can be shown that this ordering of the moments will continue when the temperature is raised, though with increasing deviations from perfect alignment, until a critical temperature is reached, above which the moments are arranged at random, as in a paramagnetic.

Weiss's theory can thus account for the fact that ferromagnetic materials may be spontaneously magnetized even in the absence of any external magnetic field; it does not, by itself, explain why the majority of ferromagnetics are not actually found in this spontaneously magnetized state, but are much more likely to have approximately zero magnetization. Weiss met this difficulty by introducing the secondary hypothesis of division into domains. He supposed that the forces of interaction only maintained the parallel alignment of the elementary moments over fairly small regions, while over longer distances it was relatively easy for the direction of alignment to change.

Weiss thus pictured a ferromagnetic material as containing a large number of 'domains' with the magnetization held constant in magnitude and direction by the interaction forces within each domain, but varying in direction from one domain to another. In this way a distinction can be made between the 'micromagnetization', or 'intrinsic magnetization', whose magnitude must everywhere be equal to the value given for the spontaneous magnetization by the interaction theory, and the 'macromagnetization', which is equal to the vector sum of the micromagnetization of all the domains and can therefore have any value from zero (when as many domains have their magnetization in one direction as in the opposite direction) to a saturation value equal to the spontaneous micromagnetization (when all domains are magnetized in the same direction). Changes in the ordinary macromagnetization are to be thought of as the result of changes in the directions of domain micromagnetizations, their magnitude remaining constant.

The domain hypothesis simplifies the problems of ferromagnetism by separating them into two almost independent parts. The first, and more fundamental, is concerned with the magnitude of the spontaneous micromagnetization; the second is concerned with its direction and the variation of this direction from place to place in a material.

This book is concerned with the second set of problems, and for the most part the magnitude of the spontaneous magnetization is accepted as a fixed quantity, without inquiry into its origin. A very brief summary of the main facts and theories will, however, be given here.

1.2. The spontaneous magnetization

The spontaneous magnetization, I_s, is the magnetization that would be observed in the absence of a magnetic field if none of the various secondary effects we shall be considering later intervened and caused the material to split up into domains with differing directions of magnetization. In practice these secondary effects always occur, and we can only observe and measure the spontaneous magnetization by applying a magnetic field strong enough to override them and cause all domain magnetizations to lie in the same direction. The spontaneous magnetization measured in this way differs slightly from the true value because the field required to align the domains will itself increase the magnetization of each domain to some extent. However, this effect is nearly always small and can be allowed for by suitable extrapolation.

The measured values of I_s are different for the different ferromagnetic metals and alloys and vary markedly with temperature for each material. The form of the temperature variation is, however, nearly the same for all materials and the results for nickel shown in fig. 1 are typical. Weiss showed that this was just the type of curve to be expected if the spontaneous magnetization was due to the interaction mechanism already indicated. He assumed that the interaction forces between the elementary magnetic moments could be represented by a field acting on each of the moments, the direction of this 'internal' or 'molecular' field being parallel to the mean magnetization and its strength proportional to the intensity of the magnetization. With this representation of the interaction forces the statistical equilibrium of an assembly of elementary moments at any given temperature can be calculated, and it is found that below a certain critical temperature (the 'Curie temperature') the assembly will be spontaneously magnetized, the intensity of magnetization varying with temperature as shown by the full curve in fig. 1. The agreement between the experimental points and the theoretical curve, calculated from such simple assumptions, is very striking. There are, moreover, various other phenomena, such as the variation of susceptibility with temperature above the Curie point and the anomaly of specific heat at the Curie point which fit into the molecular field treatment.

When the theory was first put forward, no satisfactory explanation could be given of the origin of the interaction forces producing the molecular field. There is, of course, a magnetic interaction between any two magnetic moments, but the internal field produced by this effect is only of the order of $\frac{4}{3}\pi I$, while the value required to give the curve of fig. 1 is nearly a thousand times larger. The mechanism of the interaction remained a mystery until 1928,

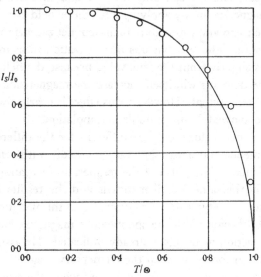

Fig. 1. Variation of spontaneous magnetization, I_s, with temperature, T. I_0 = value of I_s at absolute zero, Θ = Curie temperature. Circles show experimental values obtained by Weiss and Forrer (1926); curve obtained from Weiss's theory, modified by quantum-mechanical considerations. (See Stoner, 1948, eq. (2.11).)

when Heisenberg showed that according to the quantum theory there should be 'exchange interaction' between the different atoms that could give effects of the required type and magnitude. The Weiss molecular field theory, with quantum exchange forces as the source of the interactions, thus provides an explanation of ferromagnetism which is satisfactory in broad outline, but a more detailed examination reveals many serious discrepancies between theory and experiment, discrepancies which are often increased rather than diminished by elaboration of the theory or increased accuracy in experiment. The problem of the spontaneous magnetization is therefore by no means completely solved; the state of the theory has been reviewed recently by Stoner (1948) and by van Vleck (1945).

It has recently been shown (Néel, 1948c) that spontaneous magnetization may arise not only through positive interactions, i.e. ones tending to make neighbouring elementary moments parallel to each other, but also through negative interactions, tending to make neighbouring moments antiparallel. In a simple cubic lattice such interactions would, of course, lead to exact compensation between neighbouring antiparallel moments and so to zero spontaneous magnetization, but in more complicated lattices the attempt to make each moment antiparallel to as many as possible of its neighbours can sometimes lead to an uncompensated residue of moments all pointing in the same direction and so producing a spontaneous magnetization. It appears that the magnetism of many oxides and, in particular, of the ferrites ($Fe_2O_3 . MO$, where M is a bivalent metallic ion) is of this type, to which Néel gives the name 'ferrimagnetism'. Although the source of the spontaneous magnetization may be different, the considerations affecting the occurrence and behaviour of domains in these materials must be the same as those in ordinary ferromagnetic metals and alloys.

1.3. The directions of domain magnetizations

In most ferromagnetics the molecular field at temperatures well below their Curie temperatures is so large ($\sim 10^7$ oersteds for iron) compared with experimentally available magnetic fields that the magnitude of the spontaneous magnetization can be treated as independent of applied fields and dependent only on the temperature and composition of the material. All ordinary changes of magnetization and their related effects must therefore be explained in terms of changes of the direction of the spontaneous magnetization. According to the domain hypothesis this direction is not constant within the material, but varies from one domain to another. If the changes of direction were slow and gradual, then there would be no clear distinction between one domain and another and the idea of domains would not be a useful basis on which to discuss the properties of ferromagnetics. It turns out, however, that in most substances there is a strong tendency for the intrinsic magnetization to lie in one of a small number of directions (called 'easy' or 'preferred' directions), so that each domain is magnetized in one or other of these directions and separated from domains magnetized in different easy directions by relatively small transition zones where the magnetization is changing from one easy direction to the other. The width of these transition zones, the 'domain walls', will be calculated in Chapter V. It is often of the order of 1000 atom diameters (10^{-5} cm.), whereas the width of the domains themselves may be 100,000 or more atom diameters (10^{-3} cm.). The idea of domains as well-defined regions with an almost constant direction of magnetization is thus a valid one for many materials. Exceptions, where the domain walls are very broad, may occur, but they can best be dealt with after considering the simpler cases with narrow walls.

The concept of domains was originally a somewhat vague hypothesis, introduced to account for the difference between the theoretical intrinsic magnetization and the macromagnetization that was actually observed. The only property of domains was that they should be large enough to contain many atoms each, and small enough for all magnetic measurements to represent mean values over many domains. As knowledge of ferromagnetics has increased, more and more facts have been interpreted in terms of domains, so that the idea of domains has gradually become more definite. Recent theoretical work by Néel (1944a) and experiments by Williams, Bozorth and Shockley (1949) make it possible to say exactly what the size and shape of the domains are in certain materials, and even to follow in detail the changes in domain arrangements when the magnetization is changed. It is therefore possible to use the hypothesis of domains with confidence and to assume that in the laws governing their behaviour lie the clues to most of the phenomena of magnetization in low and moderate fields.

In an ordinary ferromagnetic material the domains are arranged in a complicated three-dimensional pattern, an equilibrium state determined by the action and interaction of many different forces. It is confusing to study the nature and mode of action of these forces in the common complicated cases, and we shall therefore, as far as possible, try to deal with simple cases where as few variables as possible have to be considered. Such cases can only be chosen if we have some knowledge of the types of forces that have to be taken into account, and we shall therefore give a brief, general survey of the factors determining domain arrangements, before considering the separate factors in more detail in later chapters. For some purposes it is convenient to discuss the equilibrium conditions in terms of the forces acting on the magnetizations of the domains, but more often it is simpler to discuss the contribution these forces make to the energy of the system and to find the equilibrium state by finding the conditions to make the energy a minimum. Our aim is therefore to find expressions for all those terms in the energy of a ferromagnetic which depend on the directions of magnetization of its domains. Such direction-dependent terms must arise from anisotropy either in the material itself or in the external influences acting on it. Internal anisotropies can be divided into natural anisotropy, which is due to the crystalline nature of the material, and imposed anisotropy, which is due to stresses of one sort or another which deform the crystal lattice. The chief external directional influence is that of magnetic fields, though in many cases the actual shape of the magnetic specimen also plays an important role, through the demagnetizing field it creates.

1.4. Magnetic field

The action of a magnetic field, H, on a domain magnetized to an intensity I_s in a direction making an angle θ with the field is simply to exert a couple $I_s H \sin \theta$ per unit volume on it, tending to turn the magnetization into the field direction. The corresponding contribution to the energy of the system is $-I_s H \cos \theta$. The field H in these expressions is, of course, the field actually acting on the domain; it may differ considerably from the applied external field, H_e, first because of the demagnetizing field of the specimen as a whole, secondly because of the local demagnetizing field depending on conditions at the domain boundary. We shall see later that these differences are often very important, but for the present they need not be considered.

1.5. Magnetocrystalline anisotropy

If magnetization curves (I against H) are measured for single crystals of a ferromagnetic it is found that much smaller fields are needed to magnetize the crystal in some directions (the easy directions) than in others; magnetization curves measured along the three principal crystallographic axes of an iron crystal are shown in fig. 2. On the domain hypothesis this is to be inter-

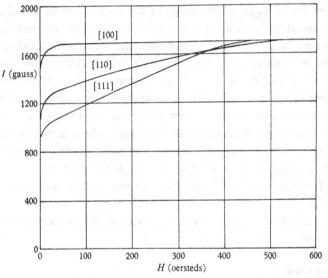

Fig. 2. Magnetization curves for the three principal axes of iron single crystals. Data from Honda and Kaya (1926).

preted as showing that there are forces tending to turn the domain magnetizations into these easy directions; or, using the energy concept, we can say that the energy of the system depends on the direction of magnetization relative to the crystal axes and that this 'magnetocrystalline' component of energy is a minimum when the magnetization lies along an easy direction. In iron there are six easy directions (counting positive and negative ones separately), those parallel to the edges of the cubic lattice. The Miller indices of these directions are [± 100], [0 ± 10] and [00 ± 1]; it is often convenient to refer to them collectively as 'the [100] directions'. In nickel, which also has a cubic lattice, the [100] directions are directions of maximum magnetocrystalline energy, and the easy directions are the eight body-diagonals, the [111] directions. Cobalt is hexagonal in structure and has two easy directions, along its principal axis ([0001] axis). The magnitude of the magneto-crystalline energy also varies from one material to another. In nickel it is of the order of 5×10^4 ergs/c.c., in iron 4×10^5 ergs/c.c. and in cobalt 5×10^6 ergs/c.c. More exact definitions and values of the magnetocrystalline energy will be given in the next chapter, where its effects are considered in detail.

1.6. Stress anisotropy

There are many materials which are approximately magnetically isotropic, either because their magnetocrystalline anisotropy is small or because they consist of many small crystals oriented at random. If, however, a stress is applied to such a material, it usually becomes anisotropic and harder to magnetize in some directions than in others. This shows that stress can produce forces affecting the directions of domain magnetization or, in other words, add to the expression for the energy of the domains terms depending on the direction of their magnetization relative to the stress directions. If, for example, a nickel wire magnetized nearly to saturation (fig. 3 a) is subjected to tension, the magnetization is found to decrease considerably, showing that the domains have turned away from the axis of the wire (fig. 3 b). In nickel under tension there is therefore a couple tending to turn the magnetization away from the axis of tension and consequently a stress-energy term depending on the direction of magnetization relative to that of tension and having its maximum value when the magnetization is parallel to the tension.

These effects are clearly connected with magnetostriction, the changes in dimensions accompanying changes in magnetization. Since the magnetization of nickel is diminished by tension, we can infer that an increase in magnetization will reduce the length, as is actually observed. The stress-energy term in the expression for domain energy, in fact, represents the work done against the stress by the magnetostrictive change in dimensions accompanying rotation of the domain magnetization. It is of order $\lambda\sigma$ per unit volume, where λ is the magnetostrictive coefficient (the fractional change in length on magnetization to saturation) and σ is the stress. λ may be negative, as in nickel, or positive, as in iron, and is not necessarily the same for all directions in a crystal.

1.7. Reasons for the occurrence of domains

In a homogeneous material subject to a uniform stress and a uniform magnetic field the three effects just considered would determine the direction of its magnetization; this direction would, however, be the same at all points in the material, so that no division into distinct domains would occur. It was originally thought that the existence of domains depended on microscopic irregulari-

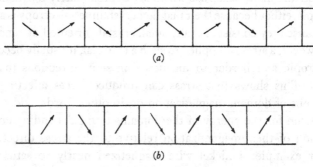

Fig. 3. Arrangement of domains in a nickel wire (schematic). (a) Moderate magnetic field acting from left to right; no stress. (b) Same magnetic field; tension applied longitudinally to wire.

ties in the crystal structure, causing the equilibrium direction of magnetization to vary from place to place, but it gradually became clear that in most cases such irregularities could play only a secondary part in the creation of domains. A more fundamental reason for their existence, even in ideally regular crystals, was pointed out by Landau and Lifshitz (1935). Any finite specimen magnetized in the same direction throughout is subject to a demagnetizing field, which can attain considerable values. A sphere of iron, for example, uniformly magnetized (fig. 4a) to its spontaneous magnetization value, I_s ($= 1720$ gauss), produces a uniform demagnetizing field of $\frac{4}{3}\pi I_s$ (about 7000 oersteds), tending to turn the magnetization into the opposite direction. This field is far larger than that required to overcome any magnetocrystalline or magnetostrictive effects which might be holding the magnetization in its original direction, so that the postulated uniformly magnetized state is clearly unstable, unless a large external magnetic field is applied. In small fields the spontaneous magnetization of the sphere must therefore be arranged in such a way as to reduce the demagnetizing field. One possible arrangement is shown in fig. 4b, where the sphere is split into laminar domains magnetized alternately in opposite directions. This arrangement removes the main demagnetizing field but leaves local fields at the edges of each domain. These fields are smaller when the domains are made thinner, but the domains cannot be multiplied indefinitely because the boundary layer between two domains has, as will be shown

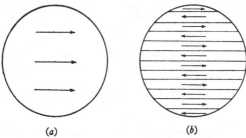

(a)   (b)

Fig. 4. Cross-sections through ferromagnetic sphere.
(a) Uniformly magnetized. (b) Split into domains.

later, a finite surface energy, and the increase of this energy has to be balanced against the decrease of energy in the demagnetizing fields as the domains are made thinner.

This effect of demagnetizing fields is more widespread than might be thought from the specialized example just considered. As is well known, the longitudinal demagnetizing effect can be made negligible by using a very long rod-shaped specimen. If we consider a long rod in which the direction of magnetization is controlled by the magnetocrystalline anisotropy, we can show that demagnetizing fields will still cause division into domains, unless the rod has a special orientation with respect to the crystal axes. Let the easy directions of magnetization defined by the anisotropy be as shown by the arrows in fig. 5a. Then the only possible uniformly magnetized arrangement that makes the magnetocrystalline energy a minimum is similar to that of fig. 5b. With this arrangement there is, however, a transverse demagnetizing field

($\sqrt{2}\,\pi I_s$ if the rod is cylindrical) tending to turn the magnetization into the easy direction B and adding a demagnetizing field energy ($\frac{1}{2}\pi I_s^2$ per unit volume of a cylinder) to the energy of the specimen. The magnetization of part of the rod will therefore turn to direction B until the average transverse field is zero and the energy reaches a minimum, giving the arrangement shown in fig. 5 c. Once again the thickness of the domains is controlled by a balance between the energy of their walls and the energy of the residual fields at the surfaces of the rod. The reduction in energy achieved by the

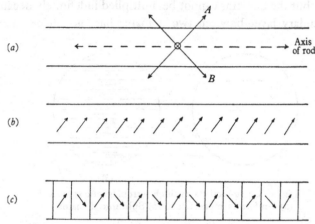

Fig. 5. Plan views of single-crystal rod. (*a*) Easy directions of magnetization (two perpendicular to plane of paper). (*b*) Uniform magnetization in direction *A*. (*c*) Domains magnetized in directions *A* and *B*.

splitting into domains is striking; an iron rod, for example, magnetized as in fig. 5 *b* would have an energy of about 10^7 ergs/c.c., which is reduced to about 100 ergs/c.c. by division into domains as in fig. 5 *c*, with the optimum thickness of domain.

The recognition of the importance of demagnetizing fields in determining domain arrangements has shown the way to produce very large domains. Williams and Shockley (1949) cut a single crystal of 3 % silicon iron into the form of a rectangular lamina with a hole through its centre (as shown in fig. 6), all the faces of the specimen being made parallel to cube-face planes of the crystal lattice. Because the easy directions of magnetization for this material are parallel to the cube edges the arrangement of domains shown in the figure is one giving minimum magnetocrystalline energy and, since there is a perfectly closed magnetic circuit, zero demagnetizing field energy. This very simple domain structure was actually observed to occur (by methods to be described in Chapter IV), the size of specimen being about $2 \times 1 \times 0.07$ cm.^{3}, so that each of the four domains had a volume of about 0.015 c.c.

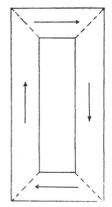

Fig. 6. Plan view of single-crystal specimen used by Williams and Shockley (1949). Arrows show directions of domain magnetizations; dotted lines indicate domain boundaries.

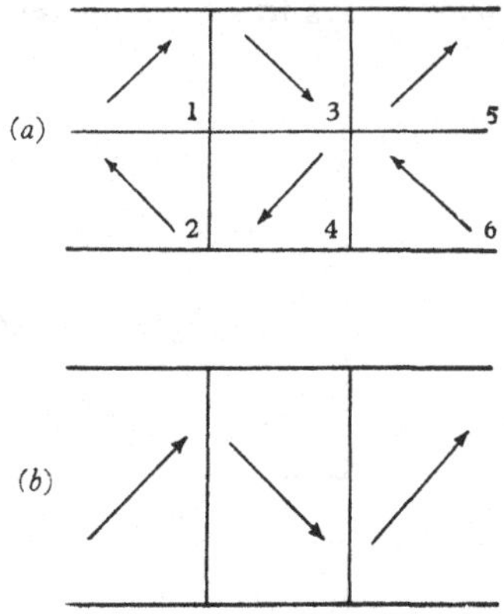

Fig. 7. Domain arrangements in a piece of a single crystal. (a) In zero field. (b) After application of a small field from left to right.

1.8. Changes in domain arrangements

We have mentioned the chief factors to be taken into account in finding the domain arrangement with lowest energy in any given circumstances, but have not yet inquired whether such an arrangement is always reached. Let fig. 7a represent the equilibrium state of a piece of a single crystal in the absence of an external field. Then the application of a field from left to right will make the state shown in fig. 7b, of lower energy than that of fig. 7a. It does not follow, however, that the change from one state to the other will actually occur in an infinitely small field. There are two ways in which the change could come about, first by the bodily rotation of the magnetization of domains 2, 4 and 6, secondly by the sideways movement of the boundary walls, leading to growth of domains 1, 3 and 5 at the expense of the others. The first method would certainly require a large field, since it involves the turning of the magnetization through a direction of maximum magnetocrystalline energy, before the new easy direction is reached. The second method might be effective for infinitely small fields if the material were perfect. It is found, however, that in all real materials there are minute imperfections, such as strained regions round an atom of impurity, and that these hinder the free movement of domain walls in various ways to be considered in detail later.

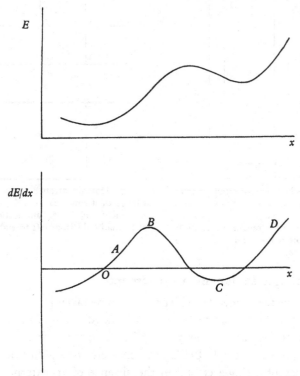

Fig. 8. Schematic representation of energy changes as a wall moves in the x direction.

It is convenient to represent the effects of these imperfections by graphs like those in fig. 8 showing the random fluctuations in the energy of the specimen as a wall moves through it, passing the various imperfections. With no applied field, the wall will be in equilibrium at a place such as O where the energy is a minimum ($dE/dx = 0$). An applied field will lower the energy of the magnetization on one side of the wall and increase it on the other, tending to displace the wall. The new equilibrium position can be found by displacing the line $dE/dx = 0$ by an amount proportional to H and finding the new point of intersection with the curve. It is clear that, for small fields, the wall will move reversibly over the region OA, but that if the field is increased enough to move the wall to B, a sudden displacement as far as D will occur, a displacement that cannot be reversed until the field is reduced again to a value corresponding to point C. The material represented in fig. 8 will therefore exhibit neither infinite permeability nor perfect reversibility of its magnetization.

1.9. Synopsis of later chapters

From the discussion of the various factors controlling domain arrangements in the last few sections, a general outline of the processes of magnetization can be built up as follows. The magnetocrystalline and magnetostrictive forces together will define two or more easy directions—directions of lowest energy of magnetization—at every point in a material. In a strain-free single crystal these directions may be the same throughout a large volume, while in material which is polycrystalline or which has irregular internal stresses they may vary rapidly from place to place. The simplest cases are those in which either the magneto-crystalline or the magnetostrictive effect is of overwhelming importance and the other can be neglected in finding the easy directions. In the absence of an external field, the domains will be arranged so that every domain is magnetized in one of the local easy directions, with the additional requirement that the general and local demagnetizing fields shall be as small as possible.

The application of a field will rearrange the domains so as to increase the magnetization in the field direction. At first this can be done, as in fig. 7, by changing domain magnetizations from one easy direction to another, without having to turn domains away from easy directions against the action of magnetocrystalline or magnetostrictive forces. At this stage the only forces opposing changes in magnetization are those which hinder the movement of domain walls as described in § 1.8. At a later stage, after the state of fig. 7b has been reached, with all domains magnetized in the easy directions nearest to the field directions, the magnetization can increase only through actual rotation of the domain magnetizations away from easy directions and closer to the field direction. This process of rotation against the action of forces of anisotropy usually requires much larger fields than those needed for the translational movement of domain walls, so that the magnetization curve often shows an abrupt 'knee', corresponding to the change from one process of magnetization to the other.

From a technical point of view the most important region of the magnetization curve of a material is that below the 'knee', but it is unfortunately just this region that is most difficult to interpret, since the processes involved depend on small irregularities in the material, usually randomly distributed, in addition to depending, like all questions of domain arrangements, on the factors determining the easy directions of magnetization.

The next two chapters of this book will therefore be concerned with the forces of magnetocrystalline anisotropy and magnetostriction in idealized homogeneous materials, ignoring, as far as possible, the effects of local variations of stress or structure. The actual arrangement of domains and the theoretical and experimental evidence for their shape and size will be discussed in Chapter IV. Then, after a chapter dealing with the properties of the boundary walls between domains, we shall consider the ways in which local irregularities can affect wall movement and so help to determine magnetic properties below the knee of the magnetization curve (Chapter VI). The final chapters will deal with the speed of magnetic changes and with the transformations of energy accompanying magnetization.

CHAPTER II

MAGNETOCRYSTALLINE ANISOTROPY

2.1. Single crystals

Magnetocrystalline anisotropy plays an important part in the behaviour of nearly all ferromagnetic materials, but its effects only appear in simple form in single-crystal specimens. Large crystals of magnetite and pyrrhotite occur naturally and were studied by Weiss (1905). This work showed that magnetic properties could depend strongly on crystal orientation, but the results were complicated by variability of composition of the materials, and little further progress was made until single crystals of the ferromagnetic metals, iron, nickel and cobalt, became available. The first work on iron was that of Beck (1918). Nickel was investigated by Sucksmith, Potter and Broadway (1928) and nickel and cobalt by Kaya (1928). Many later measurements on these metals and their alloys have been made; more detailed references are given by Bozorth (1937) and Stoner (1950).

Specimens containing large crystal grains are usually made by slow cooling through the melting-point (sometimes moving the specimen out of the furnace as it solidifies) or by the 'strain and anneal' method, where the material is stretched plastically by a small amount and then annealed at a suitable temperature (stretching by $2\frac{1}{2}\%$ and annealing at 800° C. are suitable for iron). By these methods ingots containing crystal grains several cubic centimetres in volume can be obtained (Walker, Williams and Bozorth, 1949). The crystals have to be cut out from the rest of the material and filed or ground to the desired shape. This is sometimes a matter of considerable difficulty, because severe working may cause recrystallization. The shape of the specimen is important, since the demagnetizing field often has a great influence, and it can only be calculated for certain shapes. The early workers used flat disks, a centimetre or two in diameter and a millimetre or less in thickness. Such disks can be treated as oblate spheroids, for which the demagnetizing coefficient in the plane of the disk is approximately

$\pi^2 \times$ thickness/diameter. Later workers have obtained specimens in the form of long straight rods for which the demagnetizing coefficient along the length is small and can be calculated approximately (Stoner, 1945; Osborn, 1945).

The component of magnetization of a disk at right angles to the field (I_n) can be measured quite simply by measuring the torque the field exerts on the disk. The torque is, of course, equal to HI_n. The component of magnetization parallel to the field (I_p) is usually measured by an induction method, using a ballistic galvanometer connected to a coil wound round the specimen.

2.2. Magnetization curves in directions of symmetry

In general it is found that the magnetization is not in the same direction as the applied field; if, however, the field is applied along an axis of symmetry in the crystal ([100], [110], [111] in iron and nickel; [0001], [1010] in cobalt), then the magnetization has the same direction as the field and can be represented as a function of field strength by a simple graph. Typical magnetization curves along the axes of symmetry are shown in fig. 2. The curves are measured on single crystals with very small amounts of impurities or internal stresses, so that the coercive force and hysteresis effects are small; on the scale of fig. 2 these effects, occurring in the first few oersteds of field strength, are too small to be shown. In other words, all movements of domain walls are completed in negligibly small fields, and the curves of fig. 2 represent only the effects due to turning of the domains away from easy directions as they become aligned to the field. This turning is opposed chiefly by the crystal anisotropy forces, but, because the specimens are not entirely free from internal stresses, magnetostrictive forces also play a small part in resisting the alignment of domains to the field. The rounding of the curves of fig. 2 near to the I axis is attributed to this effect of internal stresses; if magnetocrystalline forces were the only ones acting, it is assumed that the magnetization curves would meet the I axis at a definite angle.*

Such magnetization curves show clearly that certain directions in the crystal, [100] in iron, [111] in nickel, [0001] in cobalt, are easy directions, while magnetization in other directions can be con-

* But see footnote on p. 134.

siderably more difficult. The work that must be done to magnetize a specimen is proportional to the area between the magnetization curve and the I axis $\left(\int_0^I H\,dI\right)$, so that it is clear that much more energy is required to magnetize a piece of iron, say, in the [111] direction than in the [100]. Akulov (1929, 1931a) extended this idea of a direction-dependent energy of magnetization to give a formal theory of the curves of fig. 2. He suggested that the effect of magnetocrystalline forces could be represented by a 'magneto-crystalline energy' term in the expression for the free energy of the crystal. This magnetocrystalline energy would, of course, depend on the direction of the domain magnetization relative to the crystal lattice, and Akulov expressed it by a series of ascending powers of α_1, α_2, α_3, the direction cosines of the magnetization relative to the principal axes of the crystal. For cubic crystals (iron and nickel) the symmetry conditions cause many terms in the series to drop out, since the final energy expression must be independent of a change in sign of any of the α's or interchange of any two of them.

The expression for the magnetocrystalline energy, F_K, thus becomes

$$F_K = K_0 + K_1(\alpha_1^2\alpha_2^2 + \alpha_2^2\alpha_3^2 + \alpha_3^2\alpha_1^2) + K_2\alpha_1^2\alpha_2^2\alpha_3^2 + \dots \quad (2.1)$$

as far as the first two terms that are not zero or constant. K_0, K_1 and K_2 are constants characteristic of the particular material. It is, in fact, found that the term in K_2 is negligible for nickel and that terms beyond those quoted are negligible for iron. For nickel K_1 is clearly negative and for iron positive, so that the easy directions, the minima of F_K, are along [111] and [100] directions respectively. When a magnetic field is applied, an extra term is added to the free energy, the energy of the magnetization in the external field, $F_H = -I_s H \cos\theta$, where I_s is the intensity of magnetization of the domains, H is the field and θ the angle between the directions of H and I_s. The equilibrium direction of domain magnetization under the combined action of magnetocrystalline forces and an external field is such as to make the sum of the two free-energy terms a minimum, and from this condition we can deduce magnetization curves similar to those of fig. 2. We shall consider the case of iron, with H applied in a [100] direction, a [110] direction and a [111] direction.

With H in a [100] direction, the equilibrium direction for the domains is clearly along the direction of H, since F_H and F_K both have minimum values there. The magnetization in the direction of H should therefore rise to the saturation value I_s in an infinitesimal field and remain constant at that value for all higher fields.

With H in a [110] direction the domain equilibrium directions will, by symmetry, be in the (100) plane as shown in fig. 9. We can express the direction cosines, α_1, α_2, α_3, relative to the crystal axes in terms of the angle θ between H and the direction of domain magnetization I_s:

$$\alpha_1 = \cos(45° - \theta)$$
$$= \frac{1}{\sqrt{2}}(\cos\theta + \sin\theta),$$
$$\alpha_2 = \cos(45° + \theta)$$
$$= \frac{1}{\sqrt{2}}(\cos\theta - \sin\theta),$$
$$\alpha_3 = 0.$$

The expression for F_K therefore becomes

$$F_K = K_0$$
$$+ K_1(\tfrac{1}{2} - \cos^2\theta)^2 + K_2 . 0 + \dots,$$

while F_H is given by

$$F_H = -I_s H \cos\theta.$$

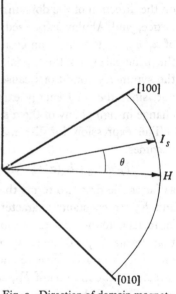

[001]

[100]

I_s

θ

H

[010]

Fig. 9. Direction of domain magnetization with a field applied to a single crystal in a [110] direction.

If we write $\cos\theta = \eta$, we get as the condition for a minimum of $F_K + F_H$

$$\frac{d(F_K + F_H)}{d\eta} = 0 = 4K_1\eta(\tfrac{1}{2} - \eta^2) + HI_s. \quad (2.2)$$

The intensity of magnetization in the direction of H is $I_s \cos\theta$, so that $\eta = I/I_s$ and the equation for the magnetization curve with H in a [110] direction is

$$\frac{HI_s}{4K_1} = \frac{I}{I_s}\left(\frac{1}{2} - \left(\frac{I}{I_s}\right)^2\right). \quad (2.3)$$

A curve drawn from this equation is shown in fig. 10 and will be seen to fit the experimental points for iron closely, assuming a suitable value for K_1.

If H is applied in a [111] direction, the domain equilibrium directions will be in (110) planes, as shown in fig. 11. The angle between the [111] direction and the [100] axis is $\cos^{-1} 1/\sqrt{3}$, so

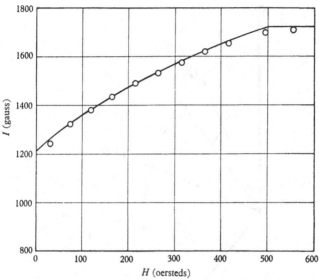

Fig. 10. Theoretical magnetization curve for iron single crystal in a [110] direction. Experimental points from Honda, Masumoto and Kaya (1928).

that we can express the direction cosines α_1, α_2, α_3 in terms of θ as follows:

$$\alpha_1 = \sqrt{\tfrac{1}{3}} \cos\theta + \sqrt{\tfrac{2}{3}} \sin\theta, \quad \alpha_2^2 = \alpha_3^2 = \tfrac{1}{2}(1 - \alpha_1^2).$$

We can thus express both F_H and F_K in terms of $\cos\theta\,(=\eta)$, and by finding the minimum of $F_H + F_K$ we obtain the rather complicated equation for the magnetization curve in the [111] direction,

$$HI_s = \frac{K_1}{3}[\eta(7\eta^2 - 3) + \sqrt{2}(4\eta^2 - 1)\sqrt{(1 - \eta^2)}]$$

$$+ \frac{K_2}{18}[\eta(1 - 16\eta^2 + 23\eta^4) - \sqrt{2}(1 - 9\eta^2 + 10\eta^4)\sqrt{(1 - \eta^2)}].$$

$$(2.4)$$

Curves drawn from this equation are shown in fig. 12. The curve neglecting the term in K_2 does not fit the experimental points so well as one taking account of the K_2 term, but both curves bend back, giving a maximum in H just before saturation is reached. This means that the curves indicate a discontinuous change in I as H is increased, a phenomenon which is not observed experimentally. It may be that consideration of higher-order terms in the

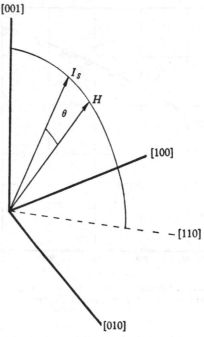

Fig. 11. Direction of domain magnetization with a field applied to a single crystal in a [111] direction.

magnetocrystalline energy would remove this anomaly, but Gans and Czerlinski (1932) have shown that its non-appearance may be due to internal stresses. They suggest that the [111] curve should be 'corrected' for internal stresses by using an experimentally observed [100] curve for similar material in the following way. Ideally, for magnetization in a [100] direction, all values of I up to I_s should be reached in zero field. In fact, because of internal stresses a finite field is required to reach any particular value of I. Gans and Czerlinski suggested that this field should be added to the field indicated by a theoretical [111] curve in order to obtain the actual field that will be required to produce the given value of I in a [111] direction. When this correction is made, the predicted and observed [111] curves agree closely.

Magnetization curves for nickel can be calculated in a similar way and also fit the experimental results well. For cobalt the expression for magnetocrystalline energy has a different form,

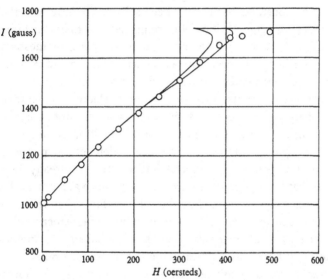

Fig. 12. Theoretical magnetization curves for iron single crystal in a [111] direction. Upper curve, neglecting K_2 term; lower curve, including K_2 term. Experimental points from Honda, Masumoto and Kaya (1928).

because the crystal is hexagonal. If the direction of domain magnetization is specified by the direction cosines $\alpha_1, \alpha_2, \alpha_3$, relative to rectangular axes, the first one coinciding with the hexagonal axis of the crystal, the lowest terms of the expansion of F_K in powers of the direction cosines are

$$K_0 + K_1 \alpha_1^2 + K_2 \alpha_1^4 + \dots,$$

all other terms up to the 4th power in the α's being either zero or constant on account of the symmetry. It is in fact found that the terms given are sufficient to represent the experimental results accurately, the properties depending only on the orientation relative to the hexagonal axis. Magnetization curves can be calculated in terms of K_1 and K_2, as for iron and nickel.

2.3. Physical origin of the anisotropy

We have introduced the quantities K as purely formal coefficients which can be used to describe the anisotropic magnetic properties of a ferromagnetic. The problem of explaining the physical origin of the anisotropy and calculating the coefficients in terms of more fundamental constants has not yet been solved. We have seen that the spontaneous magnetization of ferromagnetics is to be attributed to a coupling between elementary magnetic moments by means of quantum exchange forces. The existence of magnetocrystalline anisotropy implies that in addition to the mutual coupling between the elementary moments, now identified with electron spins, there must be coupling between these moments and the atomic lattice.

The simplest form of coupling is that resulting from purely classical magnetic interactions between elementary moments located at the points of a crystal lattice. It has already been indicated that such interactions are far too small to account for the spontaneous magnetization, but they do, in fact, give effects of the right order to account for the observed magnetocrystalline phenomena in most materials (though some substances investigated recently have anisotropies so high as to be quite outside the range that could be explained by these interactions). Unfortunately, however, the results of a considerable amount of detailed calculation by Mahajani (1929), Akulov (1929) and others is to show that the coupling provided by purely magnetic interactions cannot adequately explain the facts; in some cases it does not even provide the correct sign for the anisotropy coefficients.

The search for a more satisfactory coupling mechanism has so far proved unsuccessful, though investigations of the spin-orbit interactions (Brooks, 1940; van Vleck, 1947) show some promise. It may well prove impossible to explain the relatively feeble coupling between spins and the lattice that is responsible for the anisotropy until the far stronger coupling between one spin and another that produces the spontaneous magnetization itself is more perfectly understood. We shall not discuss further the problem of the origin of the anisotropy but shall go on to show how the formal coefficients K may be used to specify its effects in crystals with arbitrary orientation and in polycrystalline material.

2.4. Arbitrary orientation. Normal component of magnetization

We have seen that Akulov's expressions for the magnetocrystalline energy can be used to calculate magnetization curves when the field is applied along a direction of symmetry in the crystal and that the results agree well with experiment. We therefore expect that knowledge of the anisotropy constants K_1 and K_2 for a particular material should be sufficient to specify magnetocrystalline effects in it completely. The experimental results when the magnetic field is applied in a direction that is not a symmetrical one in the crystal are, however, rather complicated, and for some time it was thought that modifications of Akulov's theory would be needed to explain them. Various extensions or alterations were proposed (Takagi, 1939; von Engel and Wills, 1947), but it has recently been shown (Néel, 1944a; Lawton and Stewart, 1948) that the simple expressions already given for magnetocrystalline energy represent the facts adequately, provided that due allowance is made for some rather unusual effects of demagnetizing fields.

Experiments on single-crystal disks show that for all orientations other than ones with the field symmetrical with respect to the crystal axes there is a component of magnetization normal to the field direction (I_n), a component which can be quite large in some cases. The variation of this component with field strength and direction can be represented in several different ways, by plotting the variations of I_n as the crystal is rotated in a constant field, by plotting the locus of the end of the magnetization vector $I (= I_p + I_n,$ where I_p is the component of magnetization parallel to the applied field) as the field increases, or by plotting the variation of I_n with field for a constant field direction.

The fact that the direction of magnetization does not usually coincide with the direction of the field means that in most specimens the direction of the applied field will not be the same as that of the field actually acting in the interior of the specimen; the field in the interior is the vector sum of the applied field and the demagnetizing field, whose direction depends on the direction of the magnetization and will therefore, in general, differ from the direction of the applied field. The effect of the demagnetizing field can be seen by considering a particular case, that of a flat disk (oblate spheroid) of iron, whose plane is a (100) plane of the crystal. Four of the easy directions ([100]) will lie in the plane of the disk, and we shall suppose that a field H_e is applied in a direction making an angle ϕ with one of them.

A very large applied field will align all domains to itself so that the magnetization is I_s in the direction ϕ. The actual field acting in the crystal will be $H_e - NI_s$, where N is the demagnetizing coefficient for magnetization in the plane of the disk, but if H_e is large enough, the correction for demagnetizing field is unimportant. As H_e is reduced the domains will deviate from the direction H_e, being turned towards the easy direction [100] by the action of the magnetocrystalline forces. From the condition that the sum of magnetocrystalline energy and energy in the magnetic field should be a minimum we get the equation

$$H_e = \frac{K_1}{2I_s} \frac{\sin 4\psi}{\sin (\phi - \psi)}, \qquad (2.5)$$

where ψ is the angle between the domain magnetization and the [100] direction, so that

$$I_p = I_s \cos (\phi - \psi) \qquad (2.6)$$

and $$I_n = I_s \sin (\phi - \psi). \qquad (2.7)$$

The demagnetizing field plays no part in these equations, since it is always directed along the same line as the domain magnetization, although in the opposite direction, and so does not tend to deflect the magnetization. However, as H_e is reduced and the domains turn towards the [100] direction, this demagnetizing field will cause the true field H to deviate more and more from H_e in the opposite direction (fig. 13), until finally it lies midway between the [100] and the [010] directions. Further decrease of H_e can then no longer cause all the domains to continue moving towards the [100] direction, since this would involve the true field H moving nearer to [010] than to [100] and so cause domains to move to the easy direction near [010] rather than to that near [100]. This transfer of domains to the [010] direction will in fact occur, but only to the extent needed to maintain the resultant field H exactly midway between the [100] and [010] directions. Any deviation of the field from this symmetrical direction will cause a turning of domains from [100] to [010] or vice versa, which will change the demagnetizing field in such a way as to restore H to the symmetrical position. The range of field strengths over which H is thus maintained in the symmetrical [110] direction depends on ϕ and N, but within this range we can deduce equations for the magnetization as follows: we know that H lies in the [110] direction and that the distribution of domains between the two possible directions near [100] and near [010] is such as to produce the demagnetizing field

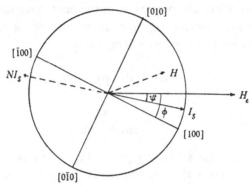

Fig. 13. Directions of fields and magnetization in a (100) single-crystal disk.

required to give H this direction; the components of H are $H_e - NI_p$ at an angle $45° - \phi$ to H, and $-NI_n$ at an angle $45° + \phi$, so that we have

$$H \cos (45° - \phi) = H_e - NI_p, \\ H \sin (45° - \phi) = NI_n, \quad (2.8)$$

or

$$I_p = \frac{H_e - H \cos (45° - \phi)}{N} \quad (2.9)$$

and

$$I_n = \frac{H}{N} \sin (45° - \phi). \quad (2.10)$$

Since H is in the [110] direction, the component of magnetization in this direction is given in terms of H by the equation already deduced for magnetization in a [110] direction (equation (2.3)). The fact that the domains are now supposed to be unequally distributed between the two easy directions does not affect the value of the magnetization along the line symmetrically between them. We can thus write H as the sum of the component of the external field in the [110] direction, $H_e \cos(45° - \phi)$, and the component of the demagnetizing field in the same direction, $-NI_{[110]}(H)$, where $I_{[110]}(H)$ is a known function defined by equation (2.3),

$$H = H_e \cos(45° - \phi) - NI_{[110]}(H). \tag{2.11}$$

Equations (2.9), (2.10) and (2.11) define the magnetization in the region where domains are distributed between two easy directions, with H midway between them. It is convenient to call this a 'two-vector' region, in distinction to the 'one-vector' region of high fields where all domains lie in the same direction. The lower limit of the two-vector region, as H_e is reduced, is reached when H in equation (2.8) becomes zero, i.e. when

$$H_e \cos(45° - \phi) - \frac{NI_s}{\sqrt{2}} = 0$$

or

$$H_e = \frac{NI_s}{\sin\phi + \cos\phi},$$

since $I_{[110]}(0) = I_s/\sqrt{2}$. From equations (2.9) and (2.10), the corresponding values of the components of magnetization are

$$I_p = \frac{H_e}{N} = \frac{I_s}{\sin\phi + \cos\phi} \tag{2.12}$$

and

$$I_n = \frac{H}{N} = 0. \tag{2.13}$$

At this point, where the actual field in the crystal is zero, the domain magnetizations lie exactly along the [100] and [010] directions, the easy directions of magnetocrystalline anisotropy. Further reduction of H_e will cause some domains to turn from these two directions to the other four easy directions to the extent required to keep the internal field equal to zero. Any departure of the field from zero will automatically be annulled by the demagnetizing field set up by the consequent turning of domains. In this lowest region of field strength, a 'six-vector' region, we have therefore the simple equations

$$I_p = H_e/N,$$

$$I_n = 0.$$

Typical magnetization curves constructed from equations (2.9)–(2.13) are shown in fig. 14. The transverse component I_n is seen to be zero in the six-vector region o—x, to increase rapidly in the two-vector region x—y, and then to fall off, as saturation is approached, in the one-vector region y—z. The calculated curves agree well with experimental ones, though the abrupt bends are

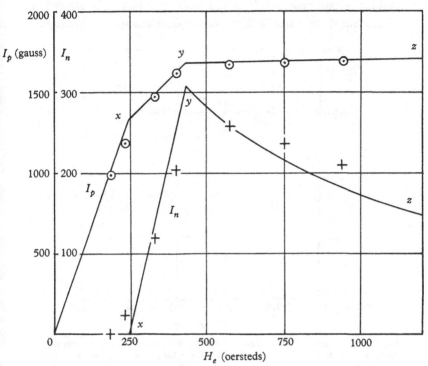

Fig. 14. Calculated magnetization curves for oblate spheroid, with $\phi = 20°$.
Experimental points from Honda and Kaya (1926).

not apparent in the latter, presumably because of imperfections in the shape or material of the crystals. The calculations just outlined for a (100) disk of iron have been extended to the case of a disk whose plane is (110) (Lawton, 1949a), and give results in accordance with experiment. Similar calculations do not appear to have been made for nickel, but there is little doubt that Akulov's expressions for magnetocrystalline energy will account for the observations on transverse magnetization in this case too, provided due allowance is made for demagnetization.

The calculation of magnetization curves for oblate spheroids has been considered in some detail because, being largely two-dimensional, it is fairly simple. A more general and complicated case is that of a long rod-shaped single crystal whose axis makes an arbitrary angle with the crystal axes. As with the oblate spheroid, the actual field, H, differs in direction from the applied field, H_e, and for certain ranges of H_e is automatically adjusted by alteration of domain distribution and the consequent changes in demagne-

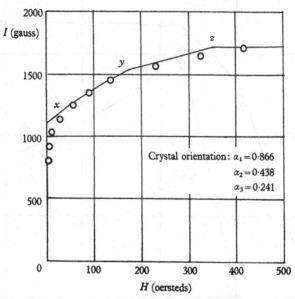

Fig. 15. Calculated magnetization curve for long rod with orientation corresponding to Sizoo's crystal 3 (1929), for which the experimental values, corrected for longitudinal demagnetizing coefficient ($N = 0·00344$), are shown as O.

tizing fields so that it has a symmetrical position among the easy directions. In iron we can distinguish four successive stages: a six-vector region where H is zero and domains are distributed between all six easy directions, a three-vector region where H is in the [111] direction nearest to the axis of the rod, a two-vector region where domains are in the two easy directions nearest to the rod axis and H is in the (110) plane between them and, finally, a one-vector region where all domains have the same direction, which approaches the axis of the rod as the field increases. Complete calculations have been carried out (Néel, 1944a; Lawton and Stewart, 1948) for rods with various particular orientations and compared with the experiments of Sizoo (1929) and Kaya (1933). A typical curve, together with experimental points, is shown in fig. 15. Experiments have been made on the component of magnetization parallel to the rod axis only; none has been made to measure the normal component in long-rod specimens. However, in order to calculate this normal component it would be necessary to know the demagnetizing coefficient perpendicular to the rod axis; for the calculation of the parallel component it is enough to assume the coefficient to be large ($\sim 2\pi$), so that the normal component of magnetization must be very small except in materials with very large K.

A characteristic feature of fig. 15 is the 'knee' at x corresponding to the transition from the six-vector to the three-vector region. At this point the domains have turned from their original random distribution among the six easy directions until they are all in one or other of the three easy directions nearest to the rod axis. The field H is, however, still zero, so that the easy directions remain along the [100] crystal axes and we also know that the component of magnetization perpendicular to the rod must be zero, in order to produce zero demagnetizing field. The component of magnetization along each easy direction is thus due to the domains lying in that easy direction only (since the other easy directions are perpendicular to it) and is equal to $f_i I_s$, where f_i is the fraction (by volume) of domains with magnetization in the easy direction i; it is also equal to $I_p \alpha_i$, where I_p is the intensity of magnetization, parallel to the rod, and α_i is the cosine of the angle between the axis of the rod and the easy direction in question. We therefore have

$$f_i I_s = I_p \alpha_i \quad (i = 1, 2, 3),$$

with $\overset{3}{\Sigma} f_i = 1$, giving

$$I_p = \frac{I_s}{\alpha_1 + \alpha_2 + \alpha_3}.$$

This simple relation between orientation and the magnetization at the knee was noticed by Kaya, and the interpretation in terms of a distribution between three easy directions in such a way as to make I_n zero was first put forward by Gorter (1933).

2.5. Polycrystalline specimens

By the methods outlined in the last section the magnetization of single crystals above the knee of their magnetization curves can be satisfactorily interpreted in terms of magnetocrystalline energy, and it is therefore interesting to inquire whether the results can be extended to include polycrystalline specimens. The first attempt in this direction was made by Gans (1932). He assumed that all the domains within each crystal grain had the same direction, namely, the direction making the sum of the magneto-crystalline energy and the energy of the domains in the external field a minimum. It was thus possible to find the component of magnetization of each grain parallel to any given external field and, by averaging over all grains, to find the mean magnetization of the polycrystalline specimen. For a specimen with its grains oriented quite at random Gans obtained expressions for the magnetization in the form of series of powers of H, giving the curve (A) shown in fig. 16.

It is clear that Gans's treatment is inadequate, except possibly in very high fields, for it assumes that each crystal grain is subject only to the external field and ignores the effects of the demagnetizing fields that will be set up by the discontinuities in magnetization at the boundaries between grains. It is nearer the truth to assume that the magnetization in each grain is very nearly the same and equal to the mean magnetization of the specimen. Any deviation from this mean value will produce divergences in the magnetization vector and give rise to demagnetizing fields that tend to restore the uniformity of magnetization. It is difficult to say with what precision this uniformity of magnetization will be maintained, since the size of demagnetizing field produced by any given deviation of the magnetization will depend on the effective demagnetizing coefficient for the particular crystal grain, a quantity determined by the shape of the grain and the properties of the surrounding grains. However, rough estimates show that in iron the magnetization should not often vary by more than a few per cent from one grain to another, at least in fields up to about 200 oersteds.

We therefore assume that each crystal grain has a component of magnetization, I, parallel to the externally applied field and a zero component normal to this direction, the local field acting on the grain being adjusted by demagnetizing effects so that it has the value needed to cause this magnetization. The local field, in fact, will be the same as the field required to cause a magnetization I in a long-rod single-crystal specimen of the same material, with the same orientation as the crystal grain in question. The mean value of these local fields for all the grains of the specimen is equal to the

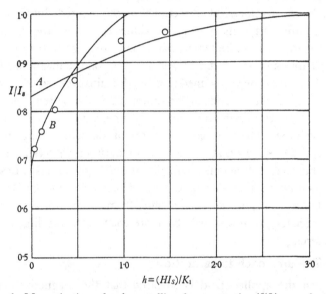

Fig. 16. Magnetization of polycrystalline ferromagnetics (I/I_s) as a function of 'reduced field', h ($=HI_s/K_1$). Curve A calculated from Gans's (1932) formulae. Curve B calculated by the method of Lawton and Stewart (1950), as explained on p. 33; circles are experimental points for the same specimen as curve B.

externally applied field, provided the specimen as a whole has a negligible demagnetizing factor in the field direction. A deviation of the local field from the external one in one grain is compensated by opposite deviations in its neighbours.

The method of constructing the magnetization curve for a polycrystal is therefore to draw curves for each of its constituent grains (or a representative sample of them) as if each grain were an infinitely long rod of the appropriate crystal orientation, and to average these curves, not by finding the mean value of I for a given

H_e, but finding the mean H_e for given I. A comparison with experiment is shown in fig. 16. The material was large-grained silicon-iron sheet, so that the grain orientations could be found by an optical method.

There are clearly various inaccuracies in this theory of magnetization in polycrystals, although fig. 16 shows that the agreement with experiment is reasonably good in fairly low fields. In the first place we saw that even the magnetization curves for single crystals in symmetrical directions were affected by internal stresses (fig. 12); in a polycrystalline specimen the internal stresses are likely to be larger, particularly near grain boundaries, and so are likely to exert a bigger influence on the magnetization. Secondly, it has been assumed that the domains were so small that the inevitable disturbance of the uniform domain pattern near the surface of a crystal would occupy only a negligible fraction of the total volume. Detailed consideration of domain arrangements (Chapter IV) shows that this is true for most single-crystal specimens but may not be true in small grains of a polycrystal. Lastly, the assumption that demagnetizing effects are large enough to maintain the magnetization the same from grain to grain is only approximately true and will be more inaccurate the higher the field strength.

2.6. The approach to saturation

When the applied field is so large that the magnetization is everywhere nearly parallel to the field, various approximations can be made, and it becomes possible to calculate the magnetization curve for polycrystals near saturation in a way that avoids the uncertainties mentioned in the last section. The general form of the 'law of approach' was indicated by Weiss (1910); if the effect of a magnetic field is to turn the magnetization vectors towards itself, the turning being resisted by internal forces, then the law of approach, when the vectors of field and magnetization are nearly parallel, must be of the form $I = I_s(1 - b/H^2)$. This follows from the equation for the equilibrium direction of the magnetization,
$$HI_s \sin \theta = C,$$
where θ is the angle between the field and the magnetization vector and C is the couple due to internal forces resisting rotation of the

magnetization. For sufficiently small angles, C can be assumed independent of θ and the magnetization in the field direction, $I_s \cos\theta$, can be written $I_s(1 - \frac{1}{2}\theta^2)$, giving the law of approach

$$I = I_s(1 - C^2/2I_s^2 H^2)$$
$$= I_s(1 - b/H^2),$$

as already stated.

If assumptions are made about the nature of the internal forces, then a value for the coefficient b can be calculated. Akulov (1931 b) was the first to calculate values of b assuming that magnetocrystalline forces were the only important ones, and his method was followed by Gans (1932) and by Becker and Döring (1939). They derived values of the coefficient b for a single crystal with any given orientation—these results have not been tested, as no measurements near enough to saturation have been made on single-crystal specimens—and, by averaging over all possible orientations, obtained a mean value which should apply to polycrystalline specimens with their grains oriented at random. The resulting law of approach for cubic crystals is

$$I = I_s\left(1 - \frac{8K_1^2}{105}\frac{1}{H^2}\right), \qquad (2.14)$$

where K_1 is the first anisotropy constant.

This law is not verified by experiment. Experimental results have usually been expressed in the form

$$I = I_s(1 + a/H + b/H^2 + ...) + \chi H, \qquad (2.15)$$

the last term representing an increase in magnetization due to increase in the intrinsic magnetization of the material. The value of χ can be calculated on the basis of the simple Weiss-Heisenberg theory; observed values are several times larger than this calculated value, but Holstein and Primakoff (1940) have shown that the discrepancy can be removed by more detailed quantum-mechanical considerations. The most complete experiments near saturation are those of Weiss and Forrer (1929) and Czerlinski (1932) for iron, and of Czerlinski (1932) and Polley (1939) for nickel. In the highest fields used (c. 10,000 oersteds) all these workers found that the term a/H was the only important one. In lower fields agreement with experiment required additional terms of equation (2.15). By identifying the coefficient b thus found for equation (2.15) with the coefficient of $1/H^2$ in equation (2.14), values of K_1 have been deduced and found to agree moderately well with those found by experiments on single crystals. The method is, however, not very trustworthy, since the theory on which equation (2.14) is based is obviously not in accordance with experiments which give a term a/H in equation (2.15).

An explanation of the a/H term was given by Brown (1941) in terms of severe local stresses in the material, but this explanation has been criticized by Néel (1948a), who reinforces Weiss's demonstration that the law of approach must be of the form $I = I_s(1 - b/H^2)$ in sufficiently high fields by pointing out that a law $I = I_s(1 - a/H)$ would lead to an infinite energy at saturation. Néel attributes the discrepancies between experiment and the Akulov theory to the latter's neglect of interactions between different grains of the polycrystal, the demagnetizing effects which we have already discussed for lower ranges of field strengths. He shows that it is only when $H \gg 4\pi I_s$, i.e. in fields greater than any so far used, that these interactions become negligible and the simple Akulov theory can be applied. In fields lower than this, but still so high that all magnetizations lie very close to the field direction, exact expressions, taking account of interactions between grains, have been obtained by Néel and by Holstein and Primakoff (1941) using Fourier series methods of representing the structure of a polycrystal. They show that in the two extreme cases $H \ll 4\pi I_s$ and $H \gg 4\pi I_s$ the law of approach takes the form $I = I_s(1 - b/H^2)$, the constant b having the value given by Akulov in the high-field region but a value tending to just half of this in the low-field region. In the intermediate region the law is complicated, but can, in practice, be reduced approximately to the form

$$I = I_s(1 - a/H).$$

2.7. Measurement of the anisotropy constants

The results outlined in the last few sections show that Akulov's series expressions (equations (2.1) and (2.3)) for magnetocrystalline energy give good agreement with experiment in all cases to which they can be applied. Such deviations from theory as occur can reasonably be attributed to approximations in the calculation, as in polycrystalline material, or to the effects of internal stresses or impurities. The magnetic anisotropy constants of a ferromagnetic, the K's of equations (2.1) and (2.3), are therefore important quantities; not only do they determine its behaviour in high fields but they also affect the properties of the walls between domains, and hence, as will be shown later, influence the low-field behaviour as well. Various methods have been used for finding the values of anisotropy constants, and four main ones can be distinguished. The most direct is to find the values of the constants that give the best fit between a theoretical magnetization curve and the experimental points. The second method is to calculate the anisotropy constants from the observed work required to magnetize a specimen to saturation in different directions. The third method is to use the limiting value of the torque acting on a single crystal in very high fields. The fourth makes use of the recently discovered phenomena of ferromagnetic resonance at high frequencies.

Until recently the first method could only be applied to measurements made on single crystals in directions of symmetry of the crystal lattice, because there was no adequate theory dealing with arbitrary orientation of single crystals or with polycrystalline specimens. The method has been applied by Gans and Czerlinski (1932) to the Japanese workers' results for iron, nickel and cobalt. They made allowance for the effect of internal strains in the way indicated on p. 22. The possibility of obtaining values of the constants from measurements on polycrystals is attractive, since single crystals are difficult to obtain; various authors have derived the anisotropy constants by fitting theoretical curves for the approach to saturation of polycrystals to the experimental points (see e.g. Becker and Döring, 1939, ch. 13), but, as has just been shown, the original theoretical expressions were in error through neglect of demagnetizing interactions between crystal grains. Néel has shown that his more complete theory can be used to obtain a value for the anisotropy constant of nickel which agrees well with those found by other methods. The method does, however, assume that the grains of the polycrystal are oriented quite at random, a condition which does not obtain in all specimens.

The second method has also been applied principally to measurements on single crystals made along principal axes of the crystal. The work done in changing the magnetization from zero to saturation is equal to $\int H \, dI$, the area between the magnetization curve and the I axis. We shall denote this work by $W_{[100]}$, $W_{[110]}$ and $W_{[111]}$ for magnetization in the three principal directions of a cubic crystal. The individual values, $W_{[100]}$, $W_{[110]}$ and $W_{[111]}$, will depend not only on the magnetocrystalline anisotropy but also on the effects of internal stresses or impurities; it is reasonable to assume, however, that these effects are isotropic, so that the differences $W_{[100]} - W_{[111]}$, etc., are determined only by the magnetocrystalline anisotropy. The difference $W_{[100]} - W_{[111]}$ is, in fact, equal to the difference in the magnetocrystalline energies (equation (2.1)) with the magnetization everywhere in the [100] direction ($\alpha_1 = 1$, $\alpha_2 = \alpha_3 = 0$) and with the magnetization everywhere in the [111] direction ($\alpha_1 = \alpha_2 = \alpha_3 = 1/\sqrt{3}$). We can thus obtain for the differences

$$W_{[100]} - W_{[111]} = -\tfrac{1}{3}K_1 - \tfrac{1}{27}K_2,$$
$$W_{[110]} - W_{[100]} = \tfrac{1}{4}K_1,$$

so that it is easy to deduce values of K_1 and K_2 from measurements of the work of magnetization in different directions. McKeehan (1937) has used this method with single crystals of various iron-nickel alloys, but since his specimens had not exactly the simple orientations assumed above, he had to elaborate the calculations by including small correcting terms. It is also possible to apply the method to find K_1 for polycrystals if it is assumed that K_2 is negligible and if the sign of K_1 is known. In the unmagnetized state all domain magnetizations will lie in easy directions ([100] if K_1 is positive, [111] if negative); in the saturated state the magnetization is everywhere in the field direction, i.e. it is equally distributed among all possible crystal directions when all the grains are taken into account. By averaging the magnetocrystalline energy given by equation (2.1) (neglecting the K_2 term) we get $F_{\text{sat.}} = \tfrac{1}{5}K_1$. In the unmagnetized state we have $F_0 = 0$ for K_1 positive and $F_0 = \tfrac{1}{3}K_1$ for K_1 negative, so that the work required to saturate a randomly oriented polycrystal is

$$W = \tfrac{1}{5}K_1 \quad \text{for } K_1 \text{ positive}$$

and

$$W = -\tfrac{2}{15}K_1 \quad \text{for } K_1 \text{ negative.}$$

In this case there is no possibility of allowing for the effects of internal stresses by comparison with measurements made in a different direction, so the method gives an upper limit for the magnitude of K_1. Nevertheless, with well-annealed nickel, Néel (1948*b*) has shown that values can be obtained agreeing closely with those found by other means.

The third method uses measurement of the torque on a specimen when a field acts on it in a known crystallographic direction. The torque is equal to the field strength multiplied by the component of magnetization normal to the field direction. We have seen that this can be calculated for simple shapes of specimen, giving the somewhat complicated curves of fig. 14. However, in sufficiently high fields, simple expressions can be obtained for the torque. Equations (2.5) and (2.7), applying to the upper part of fig. 14, show that the normal component of the magnetization is inversely proportional to H_e, so that the torque in high fields tends to a maximum value, independent of H_e. Equations (2.5) and (2.7) apply only to a disk whose plane is (100), but it is easy to derive analogous expressions for other cases. The torque exerted on the specimen by the field is simply the torque required to turn the magnetization from an easy direction to its position of equilibrium in the field; in large fields this equilibrium direction will practically coincide with the field direction, and the torque can be found by differentiating equation (2.1) and using the direction cosines of the field with respect to the crystal. In general the torque depends on both K_1 and K_2, but their effects can be separated by making measurements in more than one direction; we have already seen, for instance, that the torque in a (100) plane does not depend at all on K_2. Precautions must be taken to avoid measuring torques other than those due to magnetocrystalline anisotropy. The two chief causes of spurious torques are rotational hysteresis effects, attributable to internal stresses, and irregularities of shape giving non-uniform magnetization. Both effects become smaller the higher the field. Most workers have used flat-disk specimens, sometimes rounding the corners so that they are more nearly oblate spheroids, and so have been able to make measurements in one plane only. Brukhatov and Kirensky (1937), however, used a small single-crystal sphere of nickel and so were able to measure the torque with the field in any desired direction. Experimental techniques connected with the torque method are described by Williams (1937a).

The last method is to observe the effects of magnetocrystalline forces on resonance frequency of the electron spins. It was shown by Griffiths (1946) that if the permeability of a ferromagnetic was measured by means of a field alternating at very high frequency ($\sim$46 Mc./s.) and at the same time a steady field was applied to the specimen, then the permeability passed through a sharp maximum as this steady field was varied. This could be attributed to a resonance of the electron spins of the ferromagnetic in the steady magnetic field, and values of the ratio of magnetic moment to mechanical angular momentum of the spins—the gyromagnetic ratio—could be deduced, agreeing closely with those found by other methods. In materials with strong magnetocrystalline anisotropy the restoring force of the resonance phenomenon is not due solely to the applied steady magnetic field, but depends also on the magnetocrystalline forces. If the frequency is varied it is, in fact, possible to obtain resonance with no steady applied field at all, the spins being constrained by the magnetocrystalline forces alone. These effects have not yet been widely used to determine anisotropy constants, but Bickford (1950) has used the method to find the constant K_1 in magnetite single crystals at various temperatures.

2.8. Values of the anisotropy constants

Many measurements of the constants for various metals and alloys have been made by the methods just described. Results by different workers do not agree very well; a useful summary of values obtained for the ferromagnetic metals has been given by Stoner (1950). He concludes that the most probable value for K_1 in iron is $4 \cdot 68 \pm 0 \cdot 34 \times 10^5$ ergs/c.c.; results for K_2 vary too widely to have any significance. In nickel the value of K_1 is $-0 \cdot 50 \times 10^5$ ergs/c.c. and the value of K_2 is again uncertain. For cobalt the sum $K_1 + K_2$ is about 52×10^5 ergs/c.c., with K_1/K_2 of the order of 4. All these values are quite strongly temperature-dependent. Useful summaries of values in alloys are given in the books by Bozorth (1951) and Snoek (1947).

CHAPTER III

MAGNETOSTRICTION

3.1. Introduction

Magnetostriction is the name given to the changes of size and shape which accompany the magnetization of a ferromagnetic. Such changes were first noticed by Joule in 1842 and have since been studied by many workers. It has been found that a wide and complicated variety of deformations can be caused by magnetization in different ways; the deformations are nearly always very small, so that sensitive methods of measurement must be used. The fact that magnetization causes mechanical strain implies that mechanical stress will affect the magnetization, and it is this inverse effect that gives magnetostriction phenomena their chief importance, for they cause external or internal mechanical stresses to have a profound influence on the magnetic properties of a body. No attempt will be made here to describe the whole range of magnetostrictive effects and their interpretation; the method will be to give an outline of the formal theory that has been built up, largely through the work of Akulov (1931 a) and Becker (1930), and show how it can be used to elucidate the more important experimental results.

In Becker's treatment (Becker and Döring, 1939, or, for a simplified version, Kittel, 1949 b) a formal expression is written down for the dependence of the free energy of a crystal on the direction of magnetization and on the mechanical distortion. This is analogous to, but more general than, Akulov's expression for magnetocrystalline energy given in the preceding chapter, since it refers to a distorted crystal. The equilibrium state of magnetization and distortion in given conditions of field and stress can be found by minimizing the free energy; various simplifying assumptions have to be made to obtain manageable formulae. It is clear from this treatment that magnetostriction can be regarded as a consequence of the strain-dependence of magnetocrystalline energy, and the coefficients introduced in the theory (which can be determined by comparison with experiment) define the dependence. In the simplified treatment given below we proceed at once to write down expressions for the distortion accompanying a given state of magnetization, and from these we can determine the equilibrium state for various special cases of applied stress and magnetic field. The coefficients appearing in this treatment have less fundamental significance than those of Becker's fuller treatment, though they are, of course, related to them.

3.2. Distortion of a cubic crystal by magnetization, field and stress

The distortion of a cubic crystal (the only type to be considered here) can conveniently be represented by a strain tensor A, with components A_{ik} referred to the axes of the crystal; the components of a vector (x_0, y_0, z_0) in the zero state become

$$x = x_0 + A_{11}x_0 + A_{12}y_0 + A_{13}z_0, \text{ etc.}, \tag{3.1}$$

in the deformed state. It is a little difficult to see what state should be taken as the undeformed one, since any ferromagnetic below its Curie point will be magnetized and hence distorted. Following Becker, we adopt as 'undeformed' that cubic lattice which has the same volume as the actual lattice (which will not be exactly cubic) when all the domains of the latter have their magnetizations in easy directions.

We assume that the distortion of a crystal is a function of mechanical stress, magnetic field and direction of domain magnetization, and that the distortions from these three causes are independent and superposable. The first two types of distortion can be dealt with briefly. The effects of mechanical stress on crystals are discussed in many text-books (e.g. Love, 1926) and can be specified by appropriate elastic constants, determined for iron by Goens and Schmid (1931) and Kimura and Ohno (1934), and for nickel by Bozorth et al. (1949). The effects of a magnetic field arise from the variation of intrinsic magnetization with lattice dimensions. It is convenient to deal with the magnetic moment of a given mass of the ferromagnetic (the mass occupying 1 c.c. in the undistorted state), denoted by M_s, rather than with the more usual quantity I_s, the magnetic moment per unit volume.

It is assumed that the changes in M_s produced by distortion of the lattice depend only on the total volume change and for small changes can be represented by

$$M_s = M_s^0\left(1 + \epsilon \frac{\delta V}{V}\right), \qquad (3.2)$$

where $\delta V/V$ is the volume change and ϵ a constant. The magnetic energy of unit mass of the ferromagnetic in a field H making angle θ with the direction of magnetization is

$$HM_s \cos\theta = HM_s^0\left(1 + \epsilon \frac{\delta V}{V}\right)\cos\theta, \qquad (3.3)$$

and there is thus a tendency for the material to expand or contract in order to minimize the energy. The expansion is opposed by elastic forces, and equilibrium is reached when

$$\frac{\delta V}{V} = \frac{M_s^0 \epsilon}{\kappa}H \approx \frac{I_s \epsilon}{\kappa}H, \qquad (3.4)$$

where κ is the bulk modulus of elasticity and $\cos\theta$ has been set equal to unity, as will normally be the case in the high fields necessary to produce an appreciable change in volume. For iron ϵ is found to be approximately equal to $+0.6$ at room temperature, so that a field of 10,000 oersteds produces an expansion of only about 6 parts in 10^6.

The dependence of crystal distortion on direction of magnetization can be represented by expressing the A_{ij}'s as series expansions in the direction cosines of the magnetization, as was done for the magnetocrystalline energy in the preceding chapter. Although the symmetry is lower for the distorted lattice than for an undistorted one, symmetry considerations can still be used to simplify the series expansions.

As far as the A_{ii}'s are concerned, we may note that A_{11} expresses the change in the x component of a vector $(x_0, 0, 0)$ and must, by symmetry, be independent of a change in sign of any of the α_i's (the direction cosines of the magnetization) and of interchange of α_2 and α_3. A_{ii} must therefore be a function of α_i^2 and $\alpha_j^2\alpha_k^2$ only, so that its series expansion can be written

$$A_{ii} = a\alpha_i^2 + b\alpha_i^4 + c\alpha_j^2\alpha_k^2 + d \qquad (3.5)$$

as far as terms containing the fourth powers of the α_i's (where a, b, c and d are constants). Similarly, the A_{ij}'s, which must change sign with α_i or α_j and be independent of the sign of α_k, can be written

$$A_{ij} = e\alpha_i\alpha_j + f\alpha_i\alpha_j\alpha_k^2 + g. \tag{3.6}$$

3.3. Volume and linear magnetostriction

From these general expressions for the distortion, we may derive relations for the two quantities easiest to observe in practice, the variation of volume with direction of magnetization, and the variation of length measured in a particular direction. The volume change is given by

$$V = V_0(1 + A_{11} + A_{22} + A_{33}), \tag{3.7}$$

and, substituting values of A_{ii} from equation (3.5) and regrouping terms, we have

$$V = V_0(1 + \omega_k s), \tag{3.8}$$

where ω_k is a constant and

$$s = \alpha_1^2\alpha_2^2 + \alpha_2^2\alpha_3^2 + \alpha_3^2\alpha_1^2.$$

No detailed experimental check of this equation has been made, but the magnitude of ω_k can be estimated from the observed change of volume between the unmagnetized and the saturated state, for polycrystalline specimens. In unmagnetized material the domain magnetization will everywhere lie along easy directions, giving, as average values of s, $\bar{s}_0 = 0$ for iron and $\bar{s}_0 = \frac{1}{3}$ for nickel. When the materials are saturated the magnetization must everywhere be parallel to the field direction and, if the crystals are oriented at random, we have $\bar{s}_s = \frac{1}{5}$ for both iron and nickel. The volume change from zero magnetization to saturation is thus

$$\delta V/V_0 = +\tfrac{1}{5}\omega_k \quad \text{for iron} \tag{3.9}$$

and

$$\delta V/V_0 = -\tfrac{2}{15}\omega_k \quad \text{for nickel.} \tag{3.10}$$

The measured values of $\delta V/V_0$ are about -5×10^{-7} for iron and -0.7×10^{-7} for nickel (Kornetski, 1933) (see p. 53).

The change in length in a particular direction specified by direction cosines β_1, β_2, β_3 is given by

$$\frac{\delta l}{l} = \sum_{i,k=1}^{3} A_{ik}\beta_i\beta_k. \tag{3.11}$$

The substitution of the general expressions for the A_{ij}'s (equations (3.5) and (3.6)) in this equation leads to a rather complicated expression, whose terms can be rearranged to give

$$
\begin{aligned}
\frac{\delta l}{l} = \; & h_1(\alpha_1^2\beta_1^2 + \alpha_2^2\beta_2^2 + \alpha_3^2\beta_3^2 - \tfrac{1}{3}) \\
& + h_2(2\alpha_1\alpha_2\beta_1\beta_2 + 2\alpha_2\alpha_3\beta_2\beta_3 + 2\alpha_3\alpha_1\beta_3\beta_1) \\
& + h_3 s \,(\text{for iron}) \quad \text{or} \quad + h_3(s - \tfrac{1}{3}) \;(\text{for nickel}) \\
& + h_4(\alpha_1^4\beta_1^2 + \alpha_2^4\beta_2^2 + \alpha_3^4\beta_3^2 - \tfrac{1}{3}) \\
& + h_5(2\alpha_1\alpha_2\alpha_3^2\beta_1\beta_2 + 2\alpha_2\alpha_3\alpha_1^2\beta_2\beta_3 + 2\alpha_3\alpha_1\alpha_2^2\beta_3\beta_1), \quad (3.12)
\end{aligned}
$$

where the constants h are functions of the constants a–g of equations (3.5) and (3.6) and the right-hand side has been arranged to satisfy the zero condition $\delta l/l = 0$ when the domains are equally distributed among the easy directions. The term in h_3 represents the contribution of the volume change just discussed to the change in length in a particular direction; in most cases it is very small and can be neglected in comparison with the other terms.

It is difficult to check equation (3.12) directly by experiment because of the difficulty of ascertaining when the zero state, with domains equally distributed among all easy directions, has been achieved. It is easier to work, not with the absolute values of δl, but with the changes of δl occurring when the directions of domain magnetizations change in a known way. This method has been used (Becker and Döring, 1939) to find the values of the constants h for nickel, using the data of Masiyama (1928) on the change in length of a saturated single crystal when its magnetization is rotated from a position parallel to the direction of measurement to one perpendicular to it. This change varies with the crystal orientation, and by fitting theoretical curves calculated from equation (3.12) to the experimental points the following values of the constants h can be deduced:

$$
h_1 = -24 \times 10^{-6}, \quad h_2 = -47 \times 10^{-6},
$$
$$
h_4 = -51 \times 10^{-6}, \quad h_5 = +52 \times 10^{-6}.
$$

With these values the data can be fitted quite closely, though small differences indicate that higher order terms than those of equation (3.12) have a slight effect. If only the first two terms of equation

(3.12) are used, it is still possible to fit the data moderately well
(with errors up to 20%) with

$$h_1 = -69 \times 10^{-6} \quad \text{and} \quad h_2 = -38 \times 10^{-6}.$$

The work described in § 2.4 makes it possible to determine the
distribution of domain directions for all states between the 'knee'
of the magnetization curve and saturation, so that it now appears
possible to check equation (3.12) against measurements of the
variation of linear magnetostriction with magnetization in the
region above the knee, such as those of Kaya and Takaki (1936),
though no actual comparisons of theory and experiment have yet
been made.

As already mentioned, the experimental results for nickel can
be fitted fairly well by taking only the first two terms of equation
(3.12), and in the case of iron these are the only terms that are
usually taken into account; when only these two terms are involved,
it is convenient to use equation (3.12) in a modified form. If we
consider the longitudinal magnetostriction, the change in length
measured in the direction of magnetization, i.e. with $\alpha_i = \beta_i$, we
obtain from the first two terms of equation (3.12):

$$\delta l/l = \tfrac{2}{3}h_1 + 2(h_2 - h_1)(\alpha_1^2\alpha_2^2 + \alpha_2^2\alpha_3^2 + \alpha_3^2\alpha_1^2), \qquad (3.13)$$

and thus obtain values for the saturation magnetostriction in the
directions [100] and [111]:

$$\lambda_{100} = \tfrac{2}{3}h_1 \quad \text{and} \quad \lambda_{111} = \tfrac{2}{3}h_2, \qquad (3.14)$$

and it is the coefficients λ_{100} and λ_{111} that are usually measured and
used to specify the magnetostriction.

For iron, the values deduced from the measurements of Honda
and Masiyama (1926), Webster (1925) and Kaya and Takaki (1936)
are $\lambda_{100} = +19 \times 10^{-6}$ and $\lambda_{111} = -19 \times 10^{-6}$. For nickel the values
corresponding to the values of h_1 and h_2 already quoted are
$\lambda_{100} = -46 \times 10^{-6}$ and $\lambda_{111} = -26 \times 10^{-6}$. For a polycrystalline
specimen with its grains oriented at random the longitudinal
magnetostriction at saturation can be obtained by averaging the
α_i of equation (3.12) (first two terms) in the appropriate way
to give

$$\bar{\lambda} = \frac{2\lambda_{100} + 3\lambda_{111}}{5}.$$

In the case of nickel it is a common approximation to take the magnetostriction as isotropic ($\lambda_{100}=\lambda_{111}$) and specify it by the mean value $\bar{\lambda}$; from the single-crystal measurements we obtain $\bar{\lambda} = -34 \times 10^{-6}$, while measurements on polycrystalline nickel give values ranging from -23×10^{-6} to -47×10^{-6}, with a mean of -33×10^{-6}. The wide range of values can probably be accounted for by impurities and by uncertainties about the distribution of domain magnetizations in the demagnetized state.

3.4. Magnetostrictive energy

We have shown how the distortion of a crystal depends on the direction of magnetization of its domains; there is, of course, an inverse effect in which the direction of magnetization is affected by mechanical stress. This effect is most conveniently treated by considering the contribution to the free energy of a specimen arising from magnetostrictive distortion when the specimen is in a stress field. This energy term, F_σ, has to be added to the other terms depending on the direction of magnetization—the magneto-crystalline energy F_k and the energy in the magnetic field—if the equilibrium direction of domain magnetization is to be found by minimizing the free energy.

The stress acting at a point in a material can be specified, in a way analogous to that in which distortion was specified, by a stress tensor π_{ik}. π_{ik} defines the components of force, f_k, acting across unit area of an imaginary surface inside the material according to the equations

$$f_k = \sum_i^3 \gamma_i \pi_{ik}, \tag{3.15}$$

where $(\gamma_1, \gamma_2, \gamma_3)$ is the normal to the surface. For the simple case of a uniform pressure p, we have $-p = \pi_{11} + \pi_{22} + \pi_{33}$ and $\pi_{12} = \pi_{23} = \pi_{31} = 0$; for a simple tension, σ, in a direction $(\gamma_1, \gamma_2, \gamma_3)$, we obtain $\pi_{ik} = \sigma\gamma_i\gamma_k$. The energy of the magnetostrictive distortion A_{ik} in a material subject to a stress π_{ik} is

$$F_\sigma = \sum_{i,k=1}^3 \pi_{ik} A_{ik}. \tag{3.16}$$

To find the dependence of F_σ on direction of magnetization we must substitute values of the A_{ik} from equations (3.5) and (3.6) into equation (3.16). The complete expressions are complicated, but by taking only the leading terms of equations (3.5) and (3.6) and using the abbreviations already introduced we obtain

$$F_\sigma = -\tfrac{3}{2}\lambda_{100}[\pi_{11}\alpha_1^2 + \pi_{22}\alpha_2^2 + \pi_{33}\alpha_3^2]$$
$$-3\lambda_{111}[\alpha_1\alpha_2\pi_{12} + \alpha_2\alpha_3\pi_{23} + \alpha_3\alpha_1\pi_{31}].$$

$$(3.17)$$

The most important type of stress to be dealt with is a simple tension, σ; if this is in a direction $(\gamma_1, \gamma_2, \gamma_3)$ we have $\pi_{ik} = \sigma\gamma_i\gamma_k$, and equation (3.17) becomes

$$F_\sigma = -\tfrac{3}{2}\sigma[\lambda_{100}(\alpha_1^2\gamma_1^2 + \alpha_2^2\gamma_2^2 + \alpha_3^2\gamma_3^2)$$
$$+ 2\lambda_{111}(\alpha_1\alpha_2\gamma_1\gamma_2 + \alpha_2\alpha_3\gamma_2\gamma_3 + \alpha_3\alpha_1\gamma_3\gamma_1)].$$

$$(3.18)$$

If the magnetization is in an easy direction we have, for iron, $\alpha_i = \pm 1$, $\alpha_j = \alpha_k = 0$, and for nickel $\alpha_i = \alpha_j = \alpha_k = \pm 1/\sqrt{3}$, and equation (3.18) simplifies further to

$$F_\sigma = -\tfrac{3}{2}\lambda_{100}\sigma \cos^2\theta \quad \text{for iron,} \qquad (3.19a)$$

and $$\qquad\qquad F_\sigma = -\tfrac{3}{2}\lambda_{111}\sigma \cos^2\theta \quad \text{for nickel,} \qquad (3.19b)$$

where θ is the angle between the direction of the tension and the direction of the magnetization. If the magnetostriction is isotropic $(\lambda_{100} = \lambda_{111} = \lambda)$ then equation (3.18) simplifies to a similar simple expression

$$F_\sigma = -\tfrac{3}{2}\lambda\sigma \cos^2\theta, \qquad (3.20)$$

whatever the direction of magnetization.

We are now in a position to apply the results of the formal theory, as expressed by equations (3.12) and (3.17) and the various simplified forms derived from them, to interpret some of the many experiments that have been done on magnetostriction.

3.5. Variation of magnetostriction with magnetization

Equation (3.12) should enable us to find the magnetostrictive extension in any direction if we know the values of the α_i's, i.e. the direction of magnetization at every point. In a single crystal of iron we know that the domain magnetizations are distributed among the six easy directions in the demagnetized state, that they transfer to those directions nearest to the field direction in the early stages of magnetization, up to the 'knee', and that thereafter they turn away from easy directions and towards the field direction.

So long as the domain magnetizations are in easy directions, we have everywhere one $\alpha_i = \pm 1$ and the other two zero. Equation (3.12) then simplifies to

$$\delta l/l = \tfrac{3}{2}\lambda_{100}(\overline{\cos^2\phi} - \tfrac{1}{3}), \qquad (3.21)$$

where the simplified form, equation (3.13), has been used, ϕ is the angle between the direction in which δl is measured and the direction of magnetization, and $\overline{\cos^2\phi}$ signifies the mean of $\cos^2\phi$ over all the domain directions. We see that any turning of domains into easy directions that lie closer to the direction of measurement will cause an increase in length. This is illustrated by the results of Honda and Masiyama (1926) (fig. 17) for single crystals of iron magnetized in a [100] direction. The process of magnetization is entirely one of transfer of domains from other easy directions into the one coinciding with the field, so that the length should increase continuously to λ_{100} at saturation, according to equation (3.21). Unfortunately, the form of the curve cannot be predicted, since the distribution of domains among the possible easy directions when the crystal is not saturated is unknown, so that $\overline{\cos^2\phi}$ cannot be calculated. Heisenberg (1931) assumed that the distribution was governed by probability laws and obtained the upper magnetostriction curve of fig. 17. Akulov (1931a), with the quite different assumption that '180° changes', i.e. transfers from [−100] to [100], were completed before any '90° changes' occurred, obtained the lower curve of fig. 17.

When the magnetization and measurement of length are in the [110] directions, the transfer of domains among easy directions should give an increase in length, according to equation (3.21), until the knee is reached, at $I = I_s/\sqrt{2}$. Further increase of magnetization means that domains must turn from the easy directions towards the [110] direction, and equation (3.21) no longer applies. We can find $\delta l/l$ from equation (3.12), since we know that $\beta_1 = \beta_2 = 1/\sqrt{2}$, $\beta_3 = 0$, $\alpha_1 = \cos(45° - \phi)$, where ϕ is the angle between the domain magnetization and [110], so that $I = I_s \cos\phi$. Equation (3.12) then reduces to

$$(\delta l/l)_{110} = \tfrac{1}{4}\lambda_{100} + \tfrac{3}{4}\lambda_{111}(\cos^2\phi - \sin^2\phi)$$

$$= \tfrac{1}{4}\lambda_{100} + \tfrac{3}{4}\lambda_{111}\left(2\frac{I^2}{I_s^2} - 1\right). \qquad (3.22)$$

Remembering that λ_{111} is negative for iron, we see that the turning of domains away from easy directions should cause a decrease in length. Fig. 17 shows the experimental results, which give the expected increase up to the knee followed by a decrease. The demagnetized state does not necessarily correspond

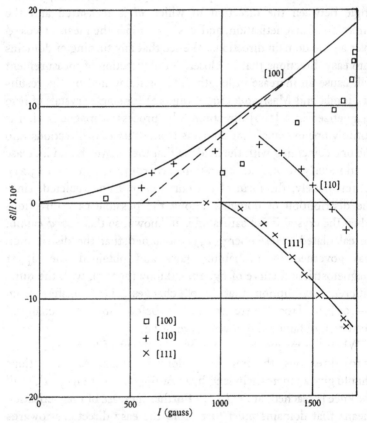

Fig. 17. Magnetostriction of iron single crystals. Theoretical curves calculated as explained in the text (upper full curve by Heisenberg (1931), broken curve by Akulov (1931)). Experimental points from Honda and Masiyama (1926).

to $\delta l = 0$, for there is no certainty that domains were equally distributed among the possible easy directions; the value of $\delta l/l$ at saturation is not well defined by experiment, so the theoretical curve has been drawn according to equation (3.22), but with a constant δl added to give the best fit to the experimental points.

Magnetostriction in the [111] direction should give no increase in length below the knee, since all six easy directions lie equally close to the [111] direction and transfers of domains do not affect equation (3.21). Above the knee the magnetostriction can be calculated from equation (3.12) to be

$$\left(\frac{\delta l}{l}\right)_{111} = \tfrac{1}{2}\lambda_{111}\left(3\frac{I^2}{I_s} - 1\right). \qquad (3.23)$$

Experimental points and a curve from equation (3.23) are shown in fig. 17.

With polycrystalline specimens of iron we expect that for magnetization below the knee there will be an increase in length as the domains of each crystallite transfer to easy directions near to the field direction. We can say little about the form of this increase, since we do not know enough about the distribution of domain directions. Above the knee we expect the length to decrease again as domains turn out of easy directions. It is possible in principle to calculate the magnetostriction curve in this region, using equation (3.12) and finding domain directions by the methods of Chapter II, but the calculation would be complicated. A simple expression can, however, be obtained for the saturation magnetostriction of a randomly oriented polycrystal. At saturation $\alpha_i = \beta_i$ for each crystal grain; putting this in equation (3.12) and averaging over all possible orientations (values of the β_i) we obtain

$$\frac{\overline{\delta l}}{l} = \bar{\lambda} = \frac{2\lambda_{100} + 3\lambda_{111}}{5}.$$

This behaviour of ordinary polycrystalline iron, giving positive magnetostriction in low fields and negative in high fields, has been known experimentally for a long time, and called, after its discoverer, the Villari reversal.

In nickel, both magnetostriction coefficients are negative and of the same order of magnitude, so that transfers of domains between easy directions and turnings away from easy directions both cause decreases in length and are hard to distinguish in magnetostriction curves. Typical results are shown in fig. 18.

When measurements are made in very high fields it is found that the magnetostriction continues to change even after the magnetization has reached saturation. This effect is due to the volume dependence of intrinsic magnetization considered on p. 43, which can be neglected for low fields in most materials but becomes important for higher ones. The effect is primarily a volume one, but an isotropic volume change $\delta V/V$ will, of course, involve a change in length $\dfrac{\delta l}{l} = \dfrac{1}{3}\dfrac{\delta V}{V}$, so that the effect appears in the measurement of longitudinal magnetostriction also. Besides the volume magnetostriction caused by volume dependence of intrinsic magnetiza-

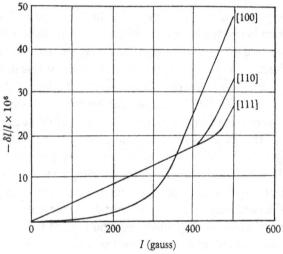

Fig. 18. Magnetostriction of nickel single crystals along principal axes. (As derived from experiments of Masiyama (1928) by Webster (1930).)

tion, that caused by volume dependence of magnetocrystalline anisotropy, expressed by equation (3.8), must also be taken into account. It has been shown (p. 44) that the magnetocrystalline effect gives a certain change in volume when the domain magnetizations change from a random distribution in the unmagnetized state to complete alignment with the field at saturation. This change should be completed in relatively low fields ($\sim$ 1000 oersteds for iron), and thereafter the intrinsic magnetization effect should give a linear increase with volume as shown by equation (3.4). Early experiments on volume magnetostriction gave results which differed widely among themselves and from the simple scheme just outlined. The difficulties were resolved in a comprehensive investigation by Kornetski (1933), who showed that they arose from neglect of another effect, the volume dependence of the energy in the demagnetizing field of a finite specimen. Many of the specimens used had been of a shape giving a large demagnetizing coefficient, N, so that the energy in the demagnetizing field, $\frac{1}{2}NVI_s^2$, was important. For specimens with very large dimension ratios, however, the shape effect is negligible, and the value for $\delta V/V$ quoted on p. 44 can be obtained. For smaller dimension ratios the observed effects agree well with the calculated ones, allowing for the demagnetizing field energy. This energy can also have an important effect on the measurement of longitudinal magnetostriction, since it can change not only through a change of the volume V but also through a change in the demagnetizing coefficient, N—an increase in length, causing N to decrease, will cause the demagnetizing field energy to diminish. There is thus a tendency for specimens with finite demagnetizing coefficients to become larger in the direction of magnetization; it is fairly easy to choose a specimen of a shape making the effect negligible.

In iron and nickel at room temperature these volume changes are small compared to the deformations produced by changes in direction of domain magnetizations and can often be neglected. Nevertheless, the volume effect is of considerable technical importance, for the great sensitivity of M_s to temperature in the region near the Curie point can lead to a decrease in volume with increasing temperature. This decrease, superimposed on the normal increase of volume with temperature, produces the very small thermal expansion coefficients associated with the Invar alloys over a certain temperature range.

3.6. Effect of stresses on magnetic properties

We have shown that the basic results for the magnetostrictive changes accompanying magnetization can be interpreted in terms of the formal theory developed by Becker. In some cases no adequate comparison between theory and experiment can be made, because too little is known about the direction of magnetization of the domains, but there is little doubt that the theory provides a satisfactory general description of the results. We must now consider the ways in which magnetostriction causes mechanical stresses to influence magnetic properties. This can best be done by reference to the free-energy function, F, that must be a minimum in the equilibrium state. We may write

$$F = F_K + F_\sigma - HI_s \cos \theta,$$

where F_K is the anisotropy energy treated in the preceding chapter, F_σ is the magnetostrictive energy given by equation (3.16) and $-HI_s \cos \theta$ is the energy in the magnetic field.

The general problem of minimizing F is complicated, but simple results can be obtained in two extreme cases, both of which can be realized approximately in practice. In the first case F_K is supposed to be negligible compared with F_σ; this is roughly true for nickel when large stresses are applied. In the second case F_K is supposed to be so large compared with F_σ that the direction of magnetization is determined almost entirely by F_K, being everywhere in one or other of the easy directions of magnetocrystalline energy; the only effect of F_σ is to make some particular easy directions slightly more favourable, from the energy point of view, than the others. We shall deal only with the simplest type of stress, a pure tension, so that F_σ is defined by equation (3.18).

For nickel we can, without introducing much error, make the further simplification of treating the magnetostriction as isotropic. The expression for F when a tension, σ, and magnetic field, H, are applied to a specimen whose magnetostriction coefficient is λ then becomes very simple:

$$F = -\tfrac{3}{2}\lambda\sigma \cos^2 \theta - HI_s \cos \theta,$$

where θ is the angle between the direction of domain magnetization and that of the field and tension. The equilibrium direction of magnetization can then be obtained from the minimum condition

$$\left.\begin{aligned} \frac{\partial F}{\partial \theta} &= 3\lambda\sigma \cos \theta \sin \theta + HI_s \sin \theta = 0, \\[2mm] \cos \theta &= -\frac{HI_s}{3\lambda\sigma}. \end{aligned}\right\} \quad (3.24)$$

or

The component of magnetization in the field direction, I, is $I_s \cos \theta$, so that the equation for the magnetization in terms of the applied field and tension becomes

$$I = -\frac{I_s^2}{3\lambda\sigma}H. \qquad (3.25)$$

The derivation of equation (3.25) can be expressed in physical terms as follows. Because of the negative magnetostriction the specimen will be longest when all its domains have their magnetizations at right angles to the axis of the specimen; this will therefore be the equilibrium arrangement when tension is applied in the absence of a magnetic field. The magnetic field H will exert

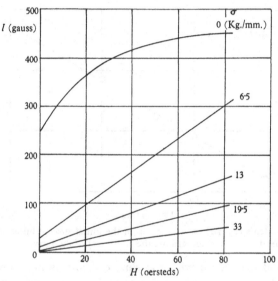

Fig. 19. Magnetization curves (descending branches of hysteresis loops) for annealed nickel wire under various tensions, σ. (After Becker and Kersten, 1930.)

a couple $HI_s \sin \theta$ on the magnetization, tending to align it to the field, and for each field strength the domain magnetizations will turn from their original positions to such an extent that this couple is just balanced by the magnetostrictive couple $\partial F_\sigma/\partial\theta$.

Fig. 19 shows the experimental results of Becker and Kersten (1930) for the magnetization of nickel with various tensions applied to it. For small tensions and fields the linear relation between I and H indicated by equation (3.25) is not obtained, but this is to be expected, because the effects of random internal strains and of the magnetocrystalline energy, which we have left out of account, will be important in this region. A graph of further results by Kersten (1931a) in fig. 20 shows the initial susceptibility, $(\partial I/\partial H)_{H=0}$, as a function of $1/\sigma$. According to equation (3.25) the susceptibility should be proportional to $1/\sigma$; fig. 20 shows that this is true and enables a value of $I_s^2/3\lambda$ to be deduced. Kersten found values from 23 to 26 Kg./mm.^{2} for different specimens. This agrees very well with the value of $I_s^2/3\lambda$ obtained by taking $I_s = 500$ and $\lambda = -3\cdot4 \times 10^{-5}$ as indicated by the experiments of Masiyama on single crystals of nickel.

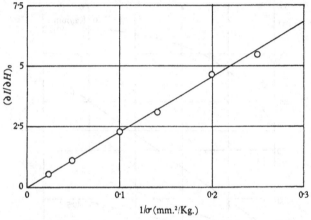

Fig. 20. Kersten's (1931a) measurements of the initial susceptibility of nickel under tension, plotted against the reciprocal of the tension.

This good agreement with the simple theory is not, however, confirmed by measurements made at higher temperatures. Fig. 21 shows the observations of Scharff (1935) for the product of susceptibility and tension, together with Döring's (1936) results for $I_s^2/3\lambda$. The two quantities are equal, as required by the theory, at room temperature, but considerable divergences appear at higher temperatures. It has recently been shown by Döring (1948a) that these divergences arise because of statistical fluctuations in I_s as the Curie point is approached, and that a more detailed treatment, using the theory of spin waves, is in agreement with experiment.

For materials with positive magnetostriction the stable solution of equation (3.24) is $\theta = 0$ or π, that is to say, for all positive values of the tension the domains all have their magnetization in line with the tension, making the length of the specimen a maximum. The action of a field H in the same direction as the tension is to turn those domains that are antiparallel to itself into the parallel position, a process which should be completed in very small fields, apart from the disturbing effects of internal stresses and magnetocrystalline anisotropy. Experimental hysteresis loops for

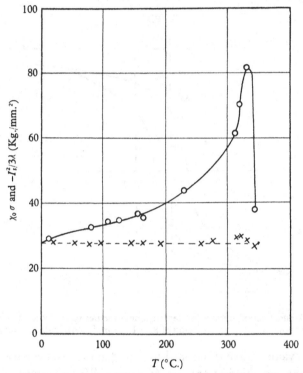

Fig. 21. Values of $\chi_0\sigma$ (Scharff, 1935), shown by full line and circles, and of $-I_s^2/3\lambda$ (Döring, 1936), shown by dotted line and crosses, plotted against temperature. (After Becker and Döring, 1939.)

Permalloy, a material with positive magnetostriction and small anisotropy, are shown in fig. 22. It is clear that with no tension there are disturbing effects which prevent perfect alignment of the domains in small fields, while when a strong enough tension is applied the domains are always constrained to lie in one of the two positions parallel to the tension; a magnetic field can cause the domains to change from one of these positions to the other, but, because of the disturbing effects, a finite field is required to bring about the change.

The observed magnetostriction is itself affected by tension. In nickel without tension, for example, the change from the demagnetized state to saturation is the result of domains changing from a random distribution among the easy directions to a state of complete alignment with the field; the corresponding change in

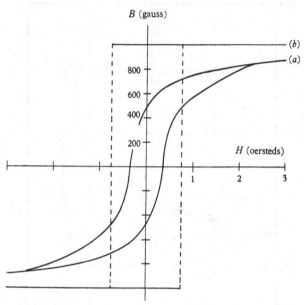

Fig. 22. Magnetization curves (hysteresis loops) for Permalloy wire. (a) With no tension applied. (b) With a tension of 8·34 Kg./mm.². (After Preisach, 1932.)

the mean value of $\cos^2 \theta$ is from $\frac{1}{3}$ to 1, so that the magnetostriction according to equation (3.20) is $\frac{3}{2}\lambda(1 - \frac{1}{3}) = \lambda$. When a large tension is applied, the initial state is one with all domains at right angles to the tension, giving $\overline{\cos^2 \theta} = 0$, and the saturated state has $\overline{\cos^2 \theta} = 1$ as before; the magnetostrictive change from the demagnetized state to saturation is therefore $\frac{3}{2}\lambda(1 - 0) = \frac{3}{2}\lambda$. Fig. 23 shows that this change of magnetostriction is observed experimentally. If a longitudinal pressure is applied to nickel (or a tension to Permalloy), then the domains should be always aligned to the pressure and the magnetostriction should vanish, changes in magnetization taking place only by 180° changes of domains, which do not affect the magnetostriction. This effect is shown in the lower curve of fig. 23.

The treatment we have just given of the effects of stresses in nickel and similar substances is based on the assumption that magnetocrystalline anisotropy and anisotropy of the magnetostriction itself can be ignored. This is justified by the agreement between theory and experiment, and by our knowledge of the

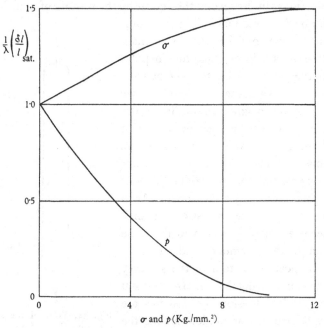

Fig. 23. Variation of magnetostriction of nickel with tension (σ) and pressure (p), from measurements by Kirchner (1936).

relative contributions of the various terms to the free-energy function F. The term F_K is of the order of the magnetocrystalline anisotropy constant, K_1, i.e. -5×10^4 ergs/c.c. for nickel; the term F_σ is of the order of $\lambda\sigma$, where $\lambda \approx 4 \times 10^{-5}$, so that a tension of about 10^9 dynes/cm.² makes the two terms equal, and larger tensions make the stress term the predominant one. In iron, on the other hand, λ is of the same order of magnitude, but K_1 is some ten times larger, and even by applying stresses up to the breaking-point of the iron it is impossible to make the stress term outweigh the magnetocrystalline one, so that the latter must always be taken

into account. An additional complication in iron is the marked
anisotropy of the magnetostriction itself. However, it is possible
to consider separately the two parts of the magnetization process in
iron and, by making appropriate approximations for each, to give a
fairly satisfactory explanation of the observed results. In the first
part, when the magnetization is less than the 'knee' value, changes
in magnetization occur by redistribution of domains among the
easy directions; in the second they occur by the turning of domains

away from easy directions. These two processes occur one after the other in the magnetization of single crystals, but in polycrystals they may overlap to some extent.

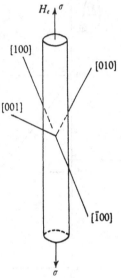

Fig. 24. Four of the six
easy directions of an iron
single-crystal rod.

In dealing with the first process we
assume that the magnetization always lies
in one or other of the easy directions of
crystal energy even when a stress acts; the
stress may, however, determine which out
of the six possible easy directions (in iron)
is actually occupied by each domain. As in
the case of nickel, the only stress we shall
consider is a uniform tension in the same
direction as the magnetic field. Fig. 24
shows the position of four of the six easy
directions in a general case. In a crystal
with no applied tension the domains are
initially distributed at random among the
easy directions; the application of a field
causes some of them to change direction
until all are in one or other of the three easy directions ([100],
[010], [001] in fig. 24) nearest to the field. As was shown in
the last chapter, it is not possible for all the domains to turn to
the [100] direction, the nearest to the field, because this would set
up a large transverse demagnetizing field. This transfer of domains
to the three most favourable easy directions, bringing the crystal
to the 'knee' of its magnetization curve, would take place in an
infinitely small field if the crystal were perfect, but in practice
there are always internal stresses or other defects which make
a finite field necessary.

When a tension σ is applied, the free energy, F, is no longer the same for all six easy directions, since the term F_σ depends on the angle the easy direction makes with the tension. If ω_1, ω_2, ω_3 are the angles the axis of the specimen makes with the crystal axes (which are the easy directions in iron), then we can, by equation $(3\cdot19a)$, write the energy differences as

$$\left.\begin{aligned}
F_{\sigma[100]} - F_{\sigma[010]} &= \tfrac{3}{2}\lambda_{100}\,\sigma(\cos^2\omega_1 - \cos^2\omega_2), \\
F_{\sigma[100]} - F_{\sigma[001]} &= \tfrac{3}{2}\lambda_{100}\,\sigma(\cos^2\omega_1 - \cos^2\omega_3).
\end{aligned}\right\} \qquad (3.26)$$

The free energy is thus a minimum when all domains are magnetized in one or other of the antiparallel pair of easy directions that lie nearest to the axis of tension. If the applied tension is greater than the internal stresses the domains will occupy these directions of minimum energy, being equally distributed between the two directions in the unmagnetized state. A field H_e parallel to the tension and the specimen axis will tend to turn the antiparallel domains from the $[-100]$ easy direction into the $[+100]$ one, and so produce a component of magnetization $I_s\cos\omega_1$ along the specimen. This process is, however, impossible, as it would also produce a component of magnetization at right angles to the specimen and so set up a large transverse demagnetizing field which would increase the free energy and make the $[+100]$ direction an unfavourable one. Arguments similar to those of the preceding chapter show that the changes of domain arrangement are closely controlled by the demagnetizing fields and that, in fact, no large transfer of domains from the $[-100]$ to the more favourable directions can occur until the external field has reached a certain critical value depending on the crystal orientation and the size of the tension σ. This critical field is one for which the internal field is large enough to compensate for the energy differences (equation (3.26)) between the [100], [010] and [001] directions, and so to enable domains to be distributed between them and to give an approximately zero transverse demagnetizing field. The form of a typical magnetization curve for an ideal single crystal of iron under tension is therefore of the form shown in fig. 25. Calculation (Stewart, 1949) shows that the slope of the initial part of the curve (OA) is very small for all orientations and is independent of the tension. The critical field corresponding to the vertical rise, BC, is given by

$$H_c = \frac{3\lambda_{100}\sigma}{2I_s}(\cos \omega_2 + \cos \omega_3)(\cos^2 \omega_1 - \sin^2 \omega_1 + \cos \omega_2 \cos \omega_3).$$

$$(3.27)$$

With tensions of 10^9 dynes/cm.^{2} the critical field, which depends on the orientation, has a maximum of about 8 oersteds for iron.

No measurements on single crystals are available to check this theory. With polycrystals we should expect the magnetization curve under tension to be a mean of curves such as that of fig. 25 over all orientations, the position of the vertical part, AB, varying widely from one crystal to another. The tension will therefore

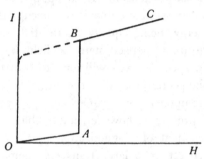

Fig. 25. Typical magnetization curve for ideal single crystal under tension (curve for same crystal without tension shown dotted).

lower the magnetization curve from its position without tension (dotted in fig. 25). The observed effect in polycrystalline iron is shown in fig. 26. The slight raising of the curve by small tensions indicates that these tensions are comparable with the random internal stresses. For larger tensions, sufficient to swamp these effects, the expected lowering is found. Experiments with polycrystals whose grains have a strong degree of preferred orientation enable rather more quantitative confirmation of the theory given above to be obtained (Stewart, 1949).

Above the knee of the magnetization curve, when domain turning movements appear, we can no longer use the simple expression, equation (3.19a), for F_σ, but must return to the full version, equation (3.18). We have therefore to find the minimum of F when both F_K and F_σ are relatively complicated, a problem for which no solution has yet been given. In this region, however,

F_K (and, of course, H) determine not only the equilibrium directions of domains but also the distribution of domains between the various possible equilibrium directions, as was shown in the preceding chapter, so that F_σ, which is small compared with F_K, cannot play more than a minor part in deciding the magnetization.

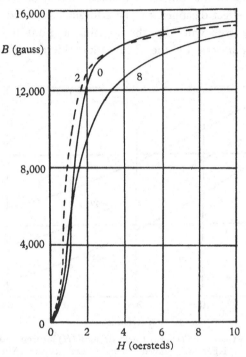

Fig. 26. Curves of flux density (B) against field strength (H) for iron under tensile stress. Tension, in Kg./mm.², shown alongside each curve. (After Bozorth and Williams, 1945.)

We can therefore make use of the thermodynamic relation

$$(\partial I/\partial\sigma)_H = (\partial l/\partial H)_\sigma \quad \text{(Stoner, 1937)}$$

to find the effect of σ on the magnetization curve from the observed magnetostrictive change in length with field. This method is of little help in the region below the knee, since $(\partial l/\partial H)_\sigma$ varies considerably with σ, but in the region where F_σ is relatively unimportant it is a good approximation to take $(\partial l/\partial H)_\sigma$ as independent of σ and therefore equal to $(\partial l/\partial H)_{\sigma=0}$. For iron this quantity is negative in the region we are considering and tension will therefore lower the magnetization curve.

The effect of tension on the magnetostriction of iron has been investigated by Honda and Shimizu (1902) and Bidwell (1890), whose results are shown in fig. 27. The increase of length with low fields is evidence of the transfer of domains to easy directions near the field. The effect disappears when a tension is applied, because the tension by itself will align the domains, and any turnings or transfers of domains by the field cannot increase the length

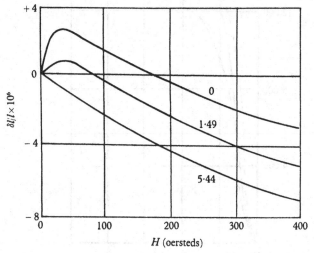

Fig. 27. Magnetostriction ($\delta l/l$) against field strength (H) for iron under tensile stress. Tension in Kg./mm.², shown alongside each curve. (After Honda and Shimizu, 1902.)

further, but may diminish it. It is clear that the material of fig. 27 had considerable internal stress, requiring fields of the order of 30 oersteds or tensions of the order of 5 Kg./mm.² to swamp them.

3.7. The 'ΔE effect'

An interesting indirect effect of magnetostriction appears in the elastic properties of ferromagnetics. If a sufficiently strong magnetic field is applied, then all the domains will be aligned to the field, and a stress will cause no change in their direction; the measured elastic properties will therefore be 'normal' ones with magnetostriction playing no part. If, however, a stress is applied to a material whose domains are free to alter their direction, then the stress may cause such an alteration, and the resultant magneto-strictive changes in length will be superimposed on the 'normal' elastic changes, giving unusual values of the elastic constants. It is fairly easy to see that the effect will always result in a larger deformation than the normal one, for a given stress, but a detailed description is complicated (Becker and Döring, 1939, ch. 24). If

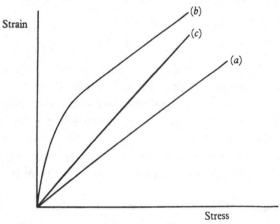

Fig. 28. Idealized form of stress-strain curves for ferromagnetic materials. (a) In large magnetic field. (b) No field, small internal stresses. (c) No field, large internal stresses.

the constraints (internal stress, magnetocrystalline forces) holding the domains in their original positions are small, then the magneto-strictive contributions to the distortion will all appear for fairly small applied stresses, so that a stress-strain curve of the type of fig. 28(b) is obtained. The straight line (a) shows the 'normal' strain curve, i.e. the one obtained in a strong magnetic field; the curve (b) shows the effect of the extra strain contribution due to magnetostriction, which appears when small stresses are applied and quickly reaches saturation with larger stresses. If there are large internal stresses present, then this saturation may not be reached until considerable stress is applied, and for small stresses the magnetostrictive contribution will increase linearly with stress, giving a stress-strain curve such as (c) in fig. 28.

S

If fig. 28 represents extension as a function of tension, then the slope of curve (a) is Young's modulus, E, for the material and the change in slope in going to curve (c) is ΔE, the change in Young's modulus due to magnetostriction. The order of magnitude of the effect can be estimated fairly easily; the saturation value of the magnetostrictive contribution, i.e. the vertical height between curves (a) and (b), will be of the order of $|\lambda|$, and if stress σ is applied to a material with internal stresses σ_i ($\sigma \ll \sigma_i$) the fraction of the saturation contribution appearing will be of order σ/σ_i. We can thus obtain a rough expression for the change in Young's modulus,

$$\frac{\Delta E}{E} \sim \frac{|\lambda|}{\sigma_i}.$$

Since it is possible to obtain independent estimates of the internal stresses from measurements of initial permeabilities (see Chapter VI), a comparison between observed and calculated ΔE effect can be made. A complete calculation of the effect must take into account the possibility of domains turning against magneto-crystalline forces, under the action of the stress. When this is done, reasonable agreement between theory and experiment is obtained.

3.8. Physical interpretation of magnetostriction

The magnetostriction effects considered in this chapter have been dealt with entirely in terms of various 'formal' coefficients, and no attempt has been made to interpret these coefficients in terms of the atomic constants of the material. Such attempts have met difficulties similar to those encountered in interpreting the magnetocrystalline coefficients and no satisfactory theory is available.

CHAPTER IV

DOMAIN ARRANGEMENT

4.1. Introduction

In the last two chapters we have considered, for the most part, simple materials such as single crystals and materials subject to uniform stresses; even in these simple materials it was found that the direction of magnetization could not, in general, be the same at all places, but that the material must be divided up in definite proportions into domains with certain definite directions of magnetization determined by the crystalline or stress anisotropy. We were able to deduce a good deal about magnetic properties by considering merely the relative volumes of domains magnetized in different directions, without inquiring into the detailed arrangement of these domains. In this chapter we shall deal with the actual size, shape and arrangement of domains in more detail. We shall consider simple materials first, with a small number of easy directions in which the domain magnetizations must lie, the same throughout the material; in more complicated materials such as polycrystals or materials with irregular stresses, where the easy directions may vary from place to place, the domain structure will clearly be more complicated and can best be elucidated after studying structures in simple materials.

4.2. Orientation of domain walls

The most important factor determining the general form of domain arrangements is the effect of the demagnetizing fields arising at the boundary between two domains. It can be shown that unless the boundary has a special orientation with respect to the directions of magnetization in the two domains it separates, demagnetizing fields will appear and will act in such a way as to shift the boundary to a position with the special orientation. The process is illustrated in a particular case by fig. 29. The wall to be considered is the portion AB separating two domains X and Y magnetized in the directions shown by the full arrows. With the wall in the position shown, the positive pole strength per unit area of wall due to the magnetization of X will be greater than the negative pole strength due to the magnetization of Y, and as a result there will be a surface divergence of magnetization on the wall which can be represented by 'free magnetic poles' spread along the wall. These poles will produce a magnetic field of the type shown by the dotted arrows, and it is clear that this field favours magnetization in the X direction near point A and in the

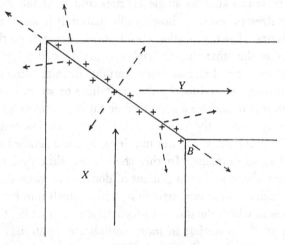

Fig. 29. Fields controlling wall orientation. Full lines represent domain boundaries; full arrows represent directions of domain magnetization; dotted arrows represent fields arising from free poles on wall AB.

Y direction near point B. There is thus a tendency for domain X to grow near point A and domain Y to grow near B so that the whole wall AB rotates. In the absence of other forces this rotation will continue until the free poles along the wall have vanished, i.e. until the wall bisects the angle between the magnetization of X and that of Y. Fig. 29 thus illustrates the general rule that the equilibrium arrangement of walls is one in which free poles are as far as possible eliminated from the interior of the material. This requirement can be expressed more formally in terms of the contribution of internal demagnetizing fields to the free energy; the total magnetic energy of a body is given by

$$W = -\int \mathbf{H}.\mathbf{I}dv + \int \frac{H_d^2}{8\pi} dv,$$

where **H** is the applied magnetic field vector, **I** the magnetization vector and H_d the demagnetizing field resulting from internal divergences of **I**. The energy will be a minimum when the second term is zero, i.e. when there are no internal demagnetizing fields, provided this can be achieved without increasing the other energy terms. The geometrical condition satisfying the requirement of no internal free poles and hence no internal demagnetizing fields is that the boundary walls should bisect the external angle between

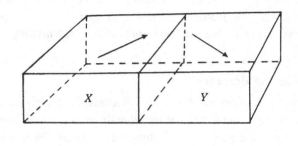

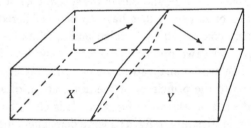

Fig. 30. Alternative wall orientations fulfilling the conditions for no free poles on the wall.

the magnetizations of the domains they separate, so that the component of magnetization normal to the wall is the same in both domains. This condition does not determine the orientation of the wall completely; it is determined as far as the plane containing the magnetization vectors of the two domains is concerned, but the angle it makes with this plane is not fixed. Fig. 30 shows two possible orientations of a wall between two domains, X and Y. We shall see later that it is usually possible to select one particular orientation as giving minimum free energy, by applying various other considerations.

A crude estimate of the energies involved in the 'demagnetizing field' mechanism for determining wall orientations can be made by considering a single plane wall turned through a small angle θ from its equilibrium position. The density of free poles on the wall (between domains with their magnetizations at right angles) will be $\sqrt{2}\,I_s \sin\theta$, and the resultant 'demagnetizing' field directed normal to the wall will be $2\pi\sqrt{2}\,I_s \sin\theta$. The magnetic energy of either domain in this field will be $2\pi I_s^2 \sin\theta$ per unit volume, or about $1\cdot8 \times 10^7 \sin\theta$ ergs/c.c. in iron. It is clear that only very small deviations from the orientation giving no free poles on the wall are possible without enormously augmenting the free energy.

4.3. Closing domains

The mechanism we have just discussed determines, in large measure, the general form of the domain arrangement, but it does little to fix the size of the individual domains. For the size of domains local demagnetizing fields are again of importance, but this time it is the fields at the boundaries of the specimen, not merely at the boundaries of domains, that have to be considered. The simplest case, first considered by Landau and Lifshitz (1935), is that of a material with two oppositely directed easy directions of magnetization. The main domain structure of such a material must consist, according to the principles of the last section, of a series of parallel-sided slabs, as shown in fig. 31 a. It is clear, however, that with the simple structure of fig. 31 a large demagnetizing fields will arise at the ends of the specimen and will tend to turn the domain magnetization near the ends away from the easy directions. It would be very difficult to determine the equilibrium configuration by formal derivation from the energy equations, but it is possible, by intuition, to select certain types of structure that clearly fulfil the equilibrium conditions better than that of fig. 31 a and to decide which of these structures gives the least free energy.

The type of structure proposed by Landau and Lifshitz for the case of fig. 31 is shown in fig. 31 b. Demagnetizing fields are entirely avoided, since the flux from domains magnetized upwards is transferred to those magnetized downwards by way of the small 'closure domains', of triangular cross-section, and does not leave the specimen at all. The closure domains are, of course, no longer magnetized in easy directions, so that they possess magneto-crystalline energy in excess of that in the bulk of the material and require a magnetic field to maintain that magnetization in an unfavourable direction. Such a field will be provided by the demagnetizing field which arises if the closure of flux is not perfect, if, for example, the sloping domain boundaries in fig. 31 b are inclined at angles slightly less than 45° to the top surface of the specimen. As we saw in the preceding section, very large fields

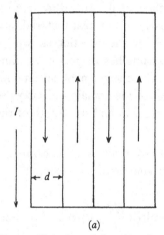
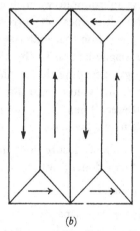

(a)   (b)

Fig. 31. Plan view of domains of a specimen with two antiparallel easy directions of magnetization. (a) Simple arrangement, with large demagnetizing fields. (b) Arrangement with 'closure domains'.

can be provided by very slight departures from the wall orientation required to give exact flux closure (no free poles on the wall) so that the actual equilibrium domain pattern will be very close to the ideal one shown in fig. 31 b, and for most materials we can assume that the conditions for perfect flux closure are met exactly. In materials where the magnetocrystalline energy (K) is large compared with the magnetic energy (I_s^2) this is no longer true, and the equilibrium structure approximates to (a) rather than (b) in fig. 31.

The extra magnetocrystalline energy, per unit volume of specimen, in the closure domains of fig. 31 b can be reduced by reducing the spacing of the main domains, and thus reducing the total volume occupied by the closure domains. It is not, however, possible to reduce the free energy of the specimen indefinitely by this means, since the reduction in volume of the closure domains is accompanied by an increase in total area of the walls between the main domains and, as will be shown in the next chapter, domain walls possess a finite surface energy (of the order of 1 erg/cm.^{2} in iron). By finding the domain spacing that makes the sum of magnetocrystalline energy in the closure domains and surface energy in the domain walls a minimum, we can determine the equilibrium size of the domains.

As an example, we may consider the case illustrated in fig. 31 b, although it does not correspond closely to any real material. We shall suppose that material magnetized in the direction shown for the magnetization in the closure domains has magnetocrystalline energy exceeding that in the rest of the specimen by an amount K per unit volume and that the energy of the domain walls is γ per unit area. Then, if the domain spacing is d, and the length of the specimen l, we have

volume of closure domains $= d/l$ per unit volume of specimen,
area of domain walls $= 1/d$ per unit volume of specimen,

and therefore

excess energy (magnetocrystalline + wall) $= \dfrac{Kd}{l} + \dfrac{\gamma}{d}$ per unit volume.

This energy will be a minimum for

$$d = \sqrt{\frac{\gamma l}{K}}. \qquad (4.1)$$

The equilibrium domain spacing therefore depends on the linear dimensions of the specimen and on the ratio of wall energy to magnetocrystalline energy. The domain structure of the simple specimen of fig. 31 is thus completely determined as to shape and size by the various considerations we have outlined.

Equation (4.1) is only valid so long as $l \gg d$. For smaller sizes of specimen it is not obvious that fig. 31 b represents the configuration with lowest energy, and the problem of finding the equilibrium structure is more difficult. For very small specimens, however, a simplification occurs, since the structure with lowest energy is one consisting of a single domain; the extra energy required for the introduction of a domain boundary is greater than the reduction in demagnetizing field energy brought about by dividing the specimen into domains. Such single-domain particles will be considered more fully in Chapter VI. In this chapter we are concerned only with specimens large enough to contain many domains. Before considering specimens more complicated and realistic than the simple one of fig. 31, it is of interest to recall briefly the various factors determining domain patterns and to note the order of magnitude of the energy associated with each, so that their relative importance may be assessed.

4.4. Survey of factors determining domain arrangement

In the first place the need to minimize the demagnetizing field of the specimen as a whole is satisfied by the division of the specimen into domains. The reduction of free energy brought about by this division is very large ($\frac{1}{2}NI_s^2$, where N is the demagnetizing coefficient, or, for an iron sphere, about 6×10^6 ergs/c.c.).

The direction of magnetization within the domains is controlled by the magnetocrystalline anisotropy or, in some materials, by magnetostrictive forces, if external or internal stresses are acting. The magnetocrystalline energy depends on the material and on its temperature, and its values range from zero up to 10^7 ergs/c.c.; in iron the change of energy when magnetization turns from an unfavourable to a favourable direction is of the order of 4×10^5 ergs/c.c.

The general shape of the domains is determined chiefly by the requirement that the density of free poles on domain boundaries shall be very low in order to avoid large demagnetizing fields near these boundaries. The excess energy near a boundary oriented in defiance of this rule would reach values of the order of 10^7 ergs/c.c. in iron.

The scale of the domain pattern is fixed by conditions at the external boundaries of the specimen, where the main domain structure usually has to be modified to avoid local demagnetizing fields. The energy of such fields depends on the details of the domain arrangements, but for a system of parallel bands of north and south poles, such as would be produced by the structure of fig. 31 a, the energy is approximately dI^2 per unit area, where d is the spacing of the bands and I the pole strength per unit area (the energy is a surface rather than a volume one because the effect of the free poles becomes negligible at distances large compared with d). Modifications of the domain pattern near the surface to reduce this demagnetizing field energy, such as those shown in fig. 31 b, usually involve small volumes of material magnetized in unfavourable directions and thus having increased anisotropy energy. It is the balance between this volume energy and the surface energy of the domain walls (of the order of 1 erg/cm.^{2} in iron) that determines the spacing of the domains.

For idealized materials, such as perfect single crystals with surfaces cut parallel to crystal planes, it is possible to find domain structures of simple form satisfying the requirements for minimum energy listed above. In real materials, which may have irregularities of external shape and internal structure, a compromise between conflicting requirements has often to be made and the actual domain structures are complicated. From the orders of magnitude of the energy terms given above it will be seen that departures from the 'ideal' orientation of domain walls require much energy and so will seldom occur over any large area of wall, but the energies involved in alterations of domain spacing are much smaller, so that quite small irregularities, such as internal stresses or surface roughnesses, may cause appreciable departures from the 'ideal' domain spacing.

4.5. Domain arrangements in a single-crystal rod

To illustrate quantitatively the various factors deciding domain structures, we shall now consider in some detail a particular example, following the treatment of Néel (1944 a). The example to be considered is a long single-crystal rod of rectangular cross-section, the crystal having cubic symmetry with 'easy directions' parallel to [100] directions (as in iron); the long axis of the rod is to be parallel to a [110] direction and the upper and lower faces of the rod parallel to the (100) plane (see fig. 32).

As was shown in Chapter II, the domain magnetizations of such a rod, in a finite magnetic field acting longitudinally, will be distributed equally between the two favourable directions nearest to the field direction; to achieve this, and at the same time to avoid the appearance of free poles on the domain boundaries, the main domains must be arranged as in fig. 33. Fig. 33a shows the arrangement in a vanishingly small field, at the 'knee' of the magnetization curve, when the domains are magnetized in the magnetocrystalline easy directions. Fig. 33b shows the change as the field is increased and domains are turned away from the easy directions; the domain boundaries remain perpendicular to the field (and to the axis of the rod).

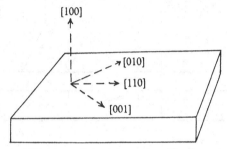

Fig. 32. Directions of crystal axes in a '[110] specimen'.

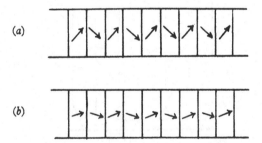

Fig. 33. Arrangement of main domains (plan view) in a [110] specimen. (a) At knee. (b) Near saturation.

In order to determine the domain spacing, we have to consider conditions at the edges of the rod and the arrangements for removing the bands of free poles that would occur with the systems of fig. 33. Various arrangements are conceivable, but Néel suggests that the ones giving minimum energy will be of the types shown in fig. 34. Although it is geometrically possible to 'close the flux' of the main domains, and thus to avoid surface free poles, with almost any orientation of the magnetic vector in the closing domains (provided these are chosen to have the right shape), Néel has shown that the orientations shown in fig. 34 are the only stable ones when the magnetocrystalline anisotropy is taken into account. The domain spacing is determined by the condition that the sum of

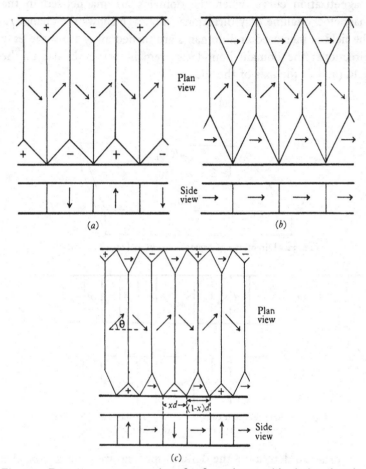

Fig. 34. Domain arrangements in a [110] specimen, with closing domains. (a), (b) The two stable arrangements of closing domains. (c) Combination of (a) and (b), giving lower energy than either.

the total wall energy in the crystal and of the excess magnetic and magnetocrystalline energy in the closing domains should be a minimum.

The free energy in the closing domains of fig. 34 is greater than that in the bulk of the material because their magnetization is not in a favourable direction. We can calculate the excess energy from equation (2.1). In the bulk of the material the energy is the sum of magnetocrystalline energy

$$F_K = \tfrac{1}{4}K_1(2\cos^2\theta - 1)^2$$

and magnetic energy $F_H = -I_s H \cos\theta$,

where θ is the angle between the direction of magnetization and that of the field, H, and is related to H by equation (2.2),

$$H = \frac{2K_1}{I_s}\cos\theta(2\cos^2\theta - 1). \qquad (4.2)$$

We therefore have

$$F = F_K + F_H = \tfrac{1}{4}K_1(2\cos^2\theta - 1)^2 - 2K_1\cos^2\theta(2\cos^2\theta - 1). \quad (4.3)$$

In closing domains of the type shown in fig. 34a, the magneto-crystalline energy is zero, since the magnetization lies along a cube edge direction, and the potential energy in the field is also to be taken as zero, since the magnetization is perpendicular to the field. The excess energy in closing domains of the first type is thus given simply by equation (4.3), and can be written

$$F_a = -F = \tfrac{1}{4}K_1(2\cos^2\theta - 1)(6\cos^2\theta + 1).$$

In closing domains of the type shown in fig. 34b, the magneto-crystalline energy is $\tfrac{1}{4}K_1$ and the magnetic energy $-HI_s$, or, by equation (4.2), $-2K_1\cos\theta(2\cos^2\theta - 1)$. The excess energy of the closing domains is thus

$$F_b = \tfrac{1}{4}K_1 - 2K_1\cos\theta(2\cos^2\theta - 1) - \tfrac{1}{4}K_1(2\cos^2\theta - 1)^2$$
$$+ 2K_1\cos^2\theta(2\cos^2\theta - 1),$$

which simplifies to

$$F_b = K_1\cos\theta(1 - \cos^2\theta)^2(2 + 3\cos\theta). \qquad (4.4)$$

The pattern with minimum energy will not, in fact, be either of those shown in fig. 34a and b, but rather a combination of the two, as shown in fig. 34c. By combining the two types of closing domain in this way it is possible to reduce their total volume without increasing the number of main domain walls. The proportion of closing domains of each type, specified by the fraction x in fig. 34c, will depend on the relative magnitudes of F_a and F_b.

Using the notation of fig. 34c we can write the excess energy in the closing domains per unit of pattern length (d) as

$$x^2 \frac{d^2}{4} \tan \theta . F_a + (1-x)^2 \frac{d^2}{4} \cot \frac{\theta}{2} . F_b.$$

If we write $W_a = \frac{1}{4} \tan \theta . F_a$ and $W_b = \frac{1}{4} \cot \frac{\theta}{2} . F_b$,

then the value of x which makes this energy expression a minimum is $x = \dfrac{W_b}{W_a + W_b}$ and the actual minimum value is $d^2 \dfrac{2W_a W_b}{W_a + W_b}$.

We can use this value for the energy in the closing domains together with values for the surface energy of the main domain walls, γ, to calculate the equilibrium value of the spacing d. The energy per unit length of the specimen is

$$W_s = 2d \frac{W_a W_b}{W_a + W_b} + \frac{2\gamma L}{d},$$

where the first term represents energy in closing domains and the second the surface energy of the main domain walls, L being the width of the specimen. Energy in the walls of the closing domains can be neglected in most cases, since their area is small. W_s has its minimum value for

$$d = \sqrt{\frac{\gamma L (W_a + W_b)}{W_a W_b}}. \qquad (4.5)$$

γ, W_a and W_b are functions of θ, so that equation (4.5) determines the domain spacing in terms of the angle through which the main magnetization has turned from the easy directions. The observed intensity of magnetization is, of course, $I = I_s \cos \theta$, and the corresponding field strength is given by equation (4.2). Table I gives numerical values for an iron crystal 1 cm. in width, and the results are plotted in fig. 35.

Fig. 35 shows the spacing becoming infinite as the field approaches zero, i.e. as the magnetization is reduced to its 'knee' value. Table I shows that this is due to the small volume energy of closing domains of type (a) in small fields; they are, in fact, magnetized along easy directions of the crystal lattice. The increase in spacing is therefore brought about by growth of closing domains of this type. Their growth would, of course, be limited before d became infinite by purely geometrical factors; if $d > L$ it

Table I. *Energies, W_a, W_b, and spacing, d, of closing domains in an iron single-crystal specimen of width L, oriented as described in the text.*

(After Néel, 1944a.)

$\cos\theta$ $=I/I_s$	H (oersteds)	W_a (ergs/c.c.)	W_b (ergs/c.c.)	$\dfrac{W_a W_b}{W_a + W_b}$ (ergs/c.c.)	γ (ergs/ cm.²)	$d/\sqrt{L}$
0·71	0	0	$6·47 \times 10^4$	0	1·21	∞
0·72	13	$0·39 \times 10^4$	6·25	$0·37 \times 10^4$	1·19	$1·79 \times 10^{-2}$
0·73	24	0·69	6·06	0·62	1·17	1·37
0·75	47	1·30	5·67	1·06	1·12	1·03
0·80	113	2·73	4·54	1·71	0·99	0·76
0·85	190	3·96	3·29	1·79	0·81	0·67
0·90	281	4·73	1·98	1·40	0·59	0·65
0·95	385	4·56	0·77	0·66	0·32	0·70

Values calculated with $K_1 = 4·2 \times 10^5$ ergs/c.c., $I_s = 1720$ gauss, using equation (4.5); wall energies, γ, calculated from equation (5.13).

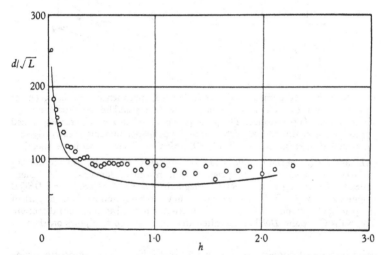

Fig. 35. Spacing of domains in a [110] single crystal as a function of 'reduced field', h. Spacing $= d$ microns, width of crystal $= L$ cm. Reduced field, $h = HI_s/K_1$. Theoretical curve from Table 1, experimental points from Bates and Neale (1950).

is impossible to construct a pattern like that of fig. 34. There is, however, another factor which will keep the closing domains much smaller than this limit. The pattern of fig. 34 is not a completely 'closed' one, for the domains of type (*a*) produce free poles on the top and bottom surfaces of the material, and in view of the high

energy associated with such poles the pattern will certainly be
modified to remove or reduce them. This can be done by adding
still more closing domains, pyramidal in shape, at the ends of the
type (*a*) domains.

When the scale of pattern is small compared with the size of
specimen these extra closing domains will not greatly affect the

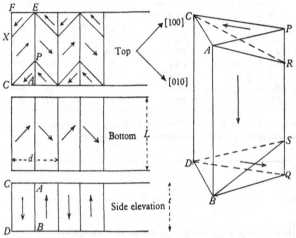

Fig. 36. Suggested domain structure for a [110] specimen in zero field. (After
Lawton, 1949.) The left-hand drawings show the top and bottom (100) surfaces,
and the vertical (110) surface. The principal domains (e.g. *PEXC*) are magnetized
in [100] or [010] directions. Fitting into the principal domains at each edge are
right-angled prisms, such as *CPABQD*, which has been drawn in perspective.
$C\hat{A}P$ and $D\hat{B}Q$ are right angles and *CAP* and *DBQ* are on the top and bottom
surfaces of the specimen. The prism is split into three domains: *CPAR* magnetized along *PC* ([$\bar{1}$oo]); *CARDSB* magnetized along *AB* ([oo$\bar{1}$)]; and *DSQB*
magnetized along *DQ* ([1oo]). *DSQB* is not a separate domain but is part of
the principal domain *PEXC*, since both are magnetized in the [100] direction.
The walls *CAR* and *DSB* are so tilted that there are no free poles on them.

calculations of spacing given above, but when the spacing becomes
comparable with the thickness of specimen they may be important.
A calculation in the general case would be complicated, but in zero
field it is easier. It can then be assumed that only small energies
are involved, so that the pattern must be accurately closed and the
magnetization of every domain must lie in an easy direction, along
a cube edge. Patterns satisfying these requirements have been
studied by Lawton (1949*b*). A possible arrangement of the closing
domain is shown in fig. 36. The balance of energies that determines

the spacing of the main pattern is no longer between wall energy in the main walls and volume energy in the closing domains, for the latter is now negligible, but between wall energy in the main walls and wall energy in the closing domains. The equilibrium spacing would be 300μ for this type of structure in an iron specimen 1 cm. wide, so that fig. 35 must be modified in the low-field region, the curve reaching $d/\sqrt{L} = 300$ at $H = 0$.

If the magnetization is reduced below the knee value, by applying a reversed field, for example, we no longer know the proportions in which the magnetization is to be divided among the various easy directions, all six of them now being available for occupation. There is therefore a wide variety of possible patterns. Walls must still be orientated in such a way as to avoid internal free poles, and the total area of walls must be restricted so that their surface energy is not too great. The general form and scale of the domain pattern will therefore be similar to those we have already discussed, with straight walls running at $45°$ to the cube axes, and forming triangular and trapezoidal domains. The actual location of individual walls will, however, depend in part on local internal stresses, which may favour different easy directions at different points, and on small surface irregularities, which may demand a special arrangement of closing domains in order to reduce the number of free poles; these small effects could be ignored while we were considering the high-field region, involving relatively high energy densities, but become important for low fields and energies.

Another effect, not considered above, which can have an important influence on the detail of domain arrangements, is the stress set up by magnetostrictive deformations around the closing domains. Rough estimates show that the effect will only be important in the region of low fields, and more precise calculations would be complicated.

4.6. Experimental study of domain arrangements: 'Bitter patterns'

The various discussions of the size and shape of domains, from the early ideas of Landau and Lifshitz to the detailed scheme of Néel (fig. 34), depend on a chain of argument and assumption so long that no great confidence in their result would be felt were it not

s   6 for the fact that it can be supported by strikingly direct experimental evidence. This evidence is obtained by applying on a small scale the familiar method of revealing magnetic lines of force by the use of iron filings. It is found that if the surface of a ferromagnetic is covered with a suspension or colloidal solution of small magnetic particles, the particles do not remain uniformly distributed but settle into a definite pattern which can be studied under a microscope. These patterns were first observed by von Hamos and Thiessen (1931) and by Bitter (1931); they have come to be known as 'Bitter patterns' or magnetic powder patterns. (Details of the experimental techniques are given by Williams,

Fig. 37. Effect of scratch on flux distribution.

Bozorth and Shockley (1949) and Mee (1950).) It has always been assumed that the patterns are due to non-uniformity of the magnetization of the material whose surface is examined, but the early experiments produced patterns so irregular that little progress could be made in interpreting them.

A great step forward was taken when Elmore (1938) emphasized the importance of careful preparation of the surface of the specimen and introduced electrolytic polishing for this purpose. Unless the surface is very smooth it is difficult to get even illumination and to distinguish the pattern, but, more important still, small ridges or hollows in the surface will cause lines of flux to emerge from the surface, as in fig. 37, so that the magnetic powder will collect at these places and the pattern will illustrate the topography rather than the magnetic state of the material. To avoid these effects the early workers polished their specimens mechanically, using emery powder and rouge or chromic oxide. This can provide a very satisfactory surface from an optical point of view, but it does so by processes involving considerable plastic flow in the surface layers. The surface layers will thus be amorphous, or at any rate polycrystalline, and will certainly have a 'crystal orientation' which is variable from place to place and differs from that of the underlying

crystals. In such layers we can clearly expect no simple domain
structure; the variations of magnetization and consequently the
powder pattern will be determined by the pattern of strains
induced by the polishing processes. Pl. I*a* is an example of such
a strain pattern; in the cubic iron lattice, slip occurs most readily
along (100) planes, so that the pattern of strains is largely a rectangular one. No detailed correlation between strains and powder
patterns appears to have been established.

Elmore avoided these strain effects by polishing his specimens
electrolytically. The surface is first made fairly smooth by
mechanical polishing and is then made the anode of an electrolytic
cell. If the electrolyte is suitable and the current density correct,
metal is removed in such a way that the remaining surface is
brightly polished, though often somewhat 'wavy'. If the elec-
trolysis is continued for long enough, the layer strained by the
mechanical polishing can be removed entirely and a surface free
from strain is obtained. Pl. I shows patterns on the same surface
after mechanical polishing (*a*) and after subsequent electrolytic
polishing (*b*). These patterns, though they are more regular, still
show little resemblance to the type predicted in fig. 34. This is not
surprising when we remember that fig. 34 was derived for a special
orientation, the upper surface being exactly a (100) plane, and
that very small deviations from this plane will produce free poles
on the surface unless the surface pattern is modified by a system
—possibly quite a complicated one—of closing domains.

A recent series of experiments by Williams, Bozorth and
Shockley (1949) has shown that by paying careful attention to the
surface orientation it is possible to avoid closing domains on the
surface and to obtain patterns showing the structure of the 'main
domains' existing in the bulk of the material. They also show how
the main domains gradually become overlaid with surface closing
domains as the orientation deviates from the ideal one. The pattern
found on a (100) plane in a crystal of 3 % silicon-iron alloy is
shown in Pl. II, together with arrows showing the directions of
magnetization in the various domains. The crystal was unmagnetized, so that all four of the easy directions in the plane of the
surface were occupied. Various triangular closing domains like
those of fig. 34 can be seen at the edges, but it is clear that their

exact size and position are largely determined by irregularities of the specimen. (The electrolytic polishing inevitably produces some rounding at the edges.)

The numerous pointed domains visible in Pl. II are closing domains of a type shown by Williams, Bozorth and Shockley to be characteristic of surfaces which deviate slightly from (100) planes.

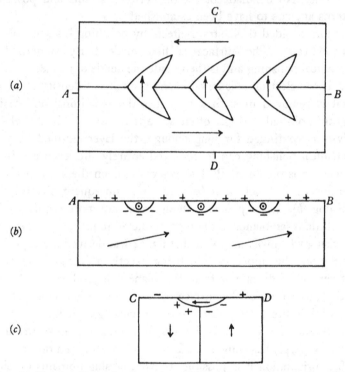

Fig. 38. Sketches illustrating formation of 'fir-tree' patterns on a surface slightly inclined to (100). Magnetization indicated by arrows, free poles (in *b* and *c*) by plus and minus signs.

Their development is shown clearly in Pl. III; the right-hand part of the surface shown was accurately parallel to a (100) plane, while the left-hand part was deliberately made to deviate from this plane by about 3°. The specimen was unmagnetized and the main domain system was one of successive slabs magnetized in opposite directions. Where the orientation is exactly (100) these give straight parallel walls and a surface clear of free poles, as on the

PLATE I

(c) Interpretation of pattern in (b); arrows show directions of domain magnetization.

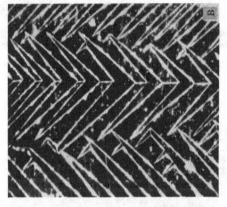

(b) Pattern on the same surface, electrolytically polished.

Crystal axes + ←——→ 0·1 mm.

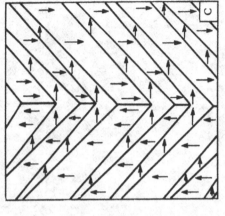

(a) Powder pattern on a mechanically polished surface.

PLATE II

Powder pattern on a (100) surface of a silicon-iron crystal.
Crystal axes +

PLATE III

Pattern showing effect of deviation from a (100) plane. Right-hand portion
is parallel to (100) plane, left-hand portion deviates by about 3°.

Crystal axes +   0·1 mm.

right of Pl. III. Where the surface deviates from (100) the main domains produce alternate bands of north and south poles, and for deviations greater than about $\frac{1}{2}°$ it is profitable to reduce the density of free poles and intermingle north and south poles more closely by the introduction of surface closing domains such as those shown in fig. 38; the numerous pointed domains visible or the left of Pl. III are clearly of this type, the so-called 'fir-tree' pattern.

Various techniques have been used to investigate the direction of magnetization in the domains revealed by the powder patterns, and they all confirm the correctness of the interpretation of the 'fir-trees' as closing domains magnetized perpendicularly to main domains. One simple method is to scratch the surface lightly in a certain direction. If the scratches are parallel to the direction of magnetization they will not disturb the flux, and little powder will collect on them; if, however, they are at right angles to the magnetization, flux will leave the surface at each scratch, as was shown in fig. 37, and powder will be attracted to the scratches.

In the original paper the energies involved in the fir-tree patterns are considered in some detail, and it is shown that their size and the amount of deviation from (100) for which they appear are in agreement with what would be expected from an approximate theoretical treatment. It is clear from many patterns such as that of Pl. II that the fir-tree type of closing domain, with its tapering sides deviating slightly from the 'no-free-pole' direction and so introducing poles of one sign into a region which would otherwise be uniformly covered with poles of the opposite sign, is one which frequently occurs. It is not confined to the simple case of oppositely magnetized slabs as in Pl. III, but also appears at the walls between domains magnetized at right angles and even in isolated form in the middle of domains, where it is presumably used to redistribute the free poles round some irregularity in the surface. Unless the surface is prepared with extreme care, the successful interpretation of powder patterns depends on distinguishing between the closing domains, which occupy only a small fraction of the volume near the surface, and the more fundamental domains forming the main magnetic structure.

The most striking example of the simplification in domain structure that can be brought about by careful attention to conditions at the surfaces has been provided by Williams and Shockley

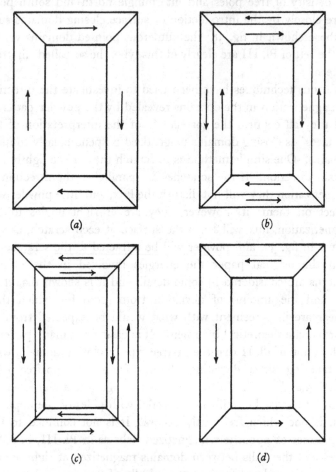

(a) (b)

(c) (d)

Fig. 39. Domain arrangements in a 'picture-frame' crystal.

(1949). They cut out a crystal of $3 \cdot 1 \%$ silicon-iron alloy in the form of a hollow rectangle about $2 \times 1 \times 0 \cdot 07$ cm.³, every one of whose faces was a (100) plane. If this is done accurately then the specimen can be magnetized to saturation in either direction round the rectangle without any closing domains at all, the structure consisting of four domains as shown in fig. 39. The change of magnetization from one direction to the other can come about by the movement of a single 180° wall across each side, intermediate positions being shown in fig. 39b–c. Powder patterns actually obtained on the specimen made it clear that the main structure was, in fact, as sketched in fig. 39, though imperfections in the surface produced many 'fir-trees'. By applying a magnetic field, Williams and Shockley were able to observe changes in the position of the wall and at the same time to measure the changes in magnetization. There was exact correlation between the two, leaving little room for doubt that the mechanism for change of magnetization was the movement of the wall under observation.

According to Williams and Shockley, the movement of the wall is not smooth as the field is varied, but proceeds in a series of jerks, the powder pattern dissolving in one place and reappearing in another. At the same time clicks indicating discontinuous changes of magnetization can be heard in a telephone and amplifier connected to a coil round the specimen. These effects are presumably due to the 'sticking' of the wall at irregularities in the crystal in a way that will be discussed later.

4.7. Comparison with theory

The patterns we have considered so far fully confirm the ideas of Landau and Lifshitz, as developed by Néel, about the importance of the condition that the number of free poles should be reduced as far as possible both on the surface and in the body of a specimen. They also support, by the general scale of the patterns, the idea that the size of domains is determined by a balance of energies in which wall surface energy plays an important part, in the general way indicated in § 4.3. It was pointed out there, however, that in the unmagnetized state the energies involved are so small that the pattern is very liable to distortion by internal stresses or surface irregularities. It is only when specimens are magnetized above the 'knee' that we can expect to get definite spacings which can be compared with those calculated by Néel.

Several authors have obtained powder-pattern lines on specimens approximating to that treated theoretically by Néel and have studied the variation of line spacing with field strength. They all confirm in a qualitative fashion the type of variation predicted by fig. 35, but in most cases the experimental conditions were too different from the theoretical ones to justify a quantitative comparison. Pl. IV shows a series of patterns obtained in increasing fields by Williams, Bozorth and Shockley on a crystal of thickness 0·18 cm., whose top surface was a (100) plane, with a width of 0·2 cm. The field was applied along the length of the specimen, which was a [110] axis. The first two pictures show the patterns obtained below the knee; the change in position of the zigzag 180° wall in the middle of the specimen, corresponding to an increase in magnetization from left to right, is clearly visible. The remaining pictures show that the domains become more closely spaced as the field increases. Néel's theory indicates that the spacing should become roughly constant in high fields, with a value of about 20μ for the size of specimen and material in question. The observed spacing is approximately 100μ. Exact agreement is not, however, to be expected, because the thickness of the crystal was so small that the free poles on the top and bottom of the type (a) domains (fig. 34) must cause serious modification of the pattern, and so the calculations giving the graph of fig. 35 do not apply.

Although they are not free from the same objection, the experiments of Bates and Neale (1950), whose results were shown in fig. 35, appear to agree more closely with theory. Bates and Mee (1952) have made measurements on a much thicker crystal ($13 \times 6 \times 6$ mm.³) which should fulfil the theoretical conditions more exactly, but the domain spacings they found agreed less well with the theory than the earlier results. They attribute this to the occurrence of more complicated closing domain structures than the simple ones shown in fig. 34 on which the calculations for fig. 35 are based. These complicated structures are probably due to slight imperfections of shape at the edges of the specimen, as is suggested by the fact that the observed domain boundaries are spaced somewhat irregularly. Confirmation of the dependence of domain spacing on the width of the specimen is provided by Pl. V, which shows a corner of a rectangular crystal with a (100) surface. As the specimen narrows towards the corner, the domains become smaller and closer together.

PLATE IV

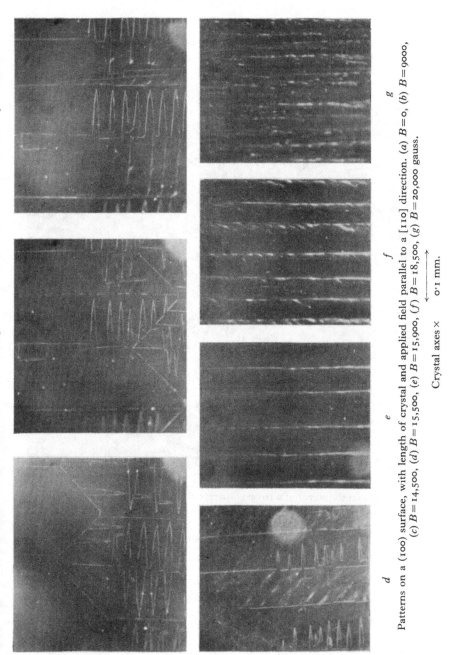

Patterns on a (100) surface, with length of crystal and applied field parallel to a [110] direction. (a) $B = 0$, (b) $B = 9000$, (c) $B = 14,500$, (d) $B = 15,500$, (e) $B = 15,900$, (f) $B = 18,500$, (g) $B = 20,000$ gauss.

Crystal axes ×   0·1 mm.

PLATE V

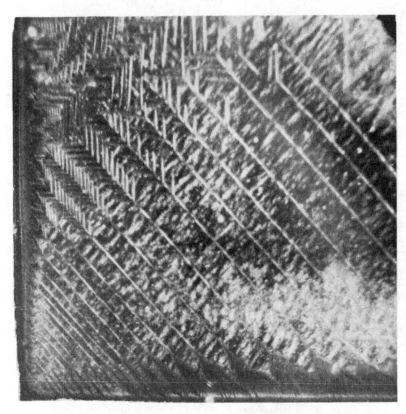

Pattern on the corner of a rectangular crystal (surface parallel to (100) plane).

Crystal axes × 0·1 mm.

The experimental results thus suggest strongly that Néel's account of the factors controlling domain size is correct in outline, though in practice it is very difficult to obtain a specimen fulfilling the conditions of the ideal case considered by him.

4.8. Mechanism of formation of powder patterns

We have not yet considered in any detail the mechanism by which the powder patterns are formed. Although it is not fully understood, and may depend on factors such as anisotropy and remanence in the particles of powder, some features of the observed patterns can be explained by using a simple model. The force on a small particle of volume v and susceptibility κ in a field H is $\frac{1}{2}\kappa v \nabla H^2$. It is convenient to split the action of the field into two parts; first it induces a magnetic moment in the particle, proportional to the field strength, then an inhomogeneous field will act on this magnetic moment and tend to move the particle to a region of maximum field strength. In powder patterns the inhomogeneous field is always provided by 'free poles' on or very near the surface of the specimen. The field chiefly responsible for inducing a magnetic moment in the powder particles may be either this local field or an external field acting over a large region, such as the field applied to the whole specimen in order to magnetize it. The former case has been considered by Williams, Bozorth and Shockley, the latter by Néel. When surface free poles provide the only fields acting on the particles, the mechanism is simply one of attraction of magnetizable particles by the poles, and the density of powder in the patterns should depend only on the density of free poles. Over most of the surface we know that this density must be small, but at the actual intersection of a domain wall with the surface there will be a high pole density. It will be shown later that the way in which the magnetization vector changes from its direction in one domain to its direction in the next, as the thickness of the wall is traversed, is by a precession about the normal to the wall, so that in the thickness of the wall there is a component of magnetization parallel to the wall and this must, in general, produce free poles where the wall intersects the surface of the specimen. We therefore expect the powder patterns with zero external field to consist of sharp lines along the actual domain boundaries, with, possibly, fainter accumulations of powder on surfaces where the density of free poles is particularly high. The patterns of Pls. I–VII are, of course, of this type; they normally take several minutes to develop after the powder suspension is applied, showing that the forces involved are small.

When a large external field is acting we have the mechanism envisaged by Néel and illustrated in fig. 40. This shows conditions at the edge of a specimen arbitrarily oriented with respect to the easy directions of domain magnetization. Alternate bands of north and south poles are produced on the surface of the specimen.

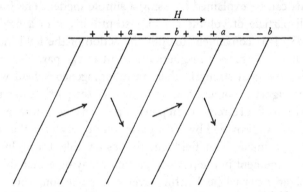

Fig. 40. Formation of powder patterns in presence of field.

Closing domains, not shown in the diagram, will, of course, be formed and will reduce the density of the free poles, but in the general case the closing domains will not remove all free poles and a system similar to that shown in fig. 40 will remain. These poles will produce fields as shown, and the powder particles, polarized by the external field, will move to the positions a, a. The alternate positions b, b are not stable ones for the particles, as can be seen by considering their polarity. The differences between this case and the zero field one are that the density of free poles spread out on surfaces between walls is greater and the density at walls (a, b) is less, while the large magnetization induced in the particles by the external field makes them more sensitive to small inhomogeneities of field but causes them to collect only in regions where the local field gradient has the appropriate sign. The practical result is that the powder lines are much less sharp than those in zero field and are formed, not at every domain boundary, but at every other one. The rate of formation of patterns is usually much greater than in the zero field case. The fact that the polarized particles can respond to rather small field gradients means that they can sometimes reveal a domain structure that is overlaid by small-scale surface closing domains or other irregularities. It is often not necessary, for instance, to resort to electrolytic polishing in order to obtain patterns in high fields. Although a mechanically polished surface certainly has no simple domain structure, the underlying main domains often cause sufficient field variations at the surface to produce a pattern when the powder suspension is applied.

The mechanism of pattern formation is discussed further in articles by Kittel (1949 a), Stoner (1950) and Mee (1950).

4.9. Powder patterns on other materials

It was pointed out earlier that iron and iron-silicon are much the most convenient materials, both experimentally and theoretically, for the study of domain structures. Work on other materials is much less complete and detailed. Until recently no satisfactory patterns have been observed on nickel, presumably because the direction of magnetization is not held firmly by magnetocrystalline forces and varies too much with local internal stresses. Bozorth and his collaborators (1950) have, however, used an alloy of 40 % nickel and 60 % iron, for which the anisotropy constant, K_1, is negative, as in nickel, but much larger (-2×10^5 ergs/c.c.), so that the easy directions are well defined along the [111] axes of the crystal. The pattern obtained on a (110) surface of such a crystal is shown in Pl. VI, together with an interpretation of the domain magnetization. The characteristic angles of 109° and 71° that would be expected if the considerations of § 4.2 are applied to a substance with [111] easy directions are clearly shown.

Patterns on nickel itself have recently been obtained by Yamamoto and Iwata (1951), Bates and Wilson (1951) and Williams and Walker (1951). They are not yet quite so clear or so easy to interpret as the patterns considered above, but their main characteristics are very similar.*

* A more complete study and interpretation of patterns on nickel has recently been published by Bates and Wilson (1953).

Patterns obtained on cobalt are shown in Pl. VII. A laminar domain structure with walls parallel to [1000] directions is clearly indicated on the 'face' surface of the crystal (Pl. VII a) but the lace-like pattern on the 'end' surface (Pl. VII b) has not yet been completely interpreted.

4.10. Other methods of studying domain structures

Although the powder-pattern method has provided by far the greatest bulk of experimental data on domain structures, two other methods have recently become available. The first (Germer, 1942; Marton, 1949) uses electron-beam techniques; electrons passing near the surface of a ferromagnetic are deflected by the stray magnetic fields at the domain boundaries and produce complex patterns on a screen or photographic plate. The connexion between the observed patterns and the domain structure is less direct than in the powder-pattern technique, and little useful information has been obtained so far. The second method (Williams, Foster and Wood, 1951; Fowler and Fryer, 1952) uses polarized light; by the Kerr effect the plane of polarization is rotated slightly on reflexion at the surface of a magnetized body, and the domain structure is revealed by differences in the rotation. The method appears to have considerable possibilities for the direct observation of domains, but it is not yet very far developed.

A method which was at one time thought to give much information about domains is the study of the 'Barkhausen effect', the small irreversible changes in magnetization which occur when ferromagnetic specimens are subjected to smoothly changing fields and which can be measured by suitable search coils and amplifiers. It has gradually become clear that these irreversible changes do not correspond to rotations of the magnetization of whole domains, as was once thought, but rather to irreversible movements of domain walls, as described on p. 87, which may affect only small parts of the domains. The effect therefore gives very little information about domain structure. Recent work by Tebble and his collaborators (1950, 1953) has shown that the effect does give a certain amount of information about the reversibility of changes of magnetization, but there are many complicating factors and the work will not be considered further here.

PLATE VI

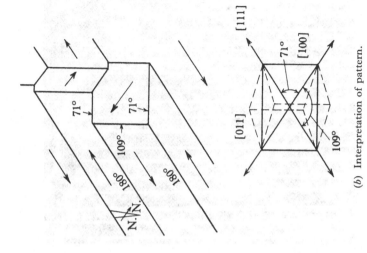

(a) Pattern obtained on a (110) surface
of a cobalt-nickel crystal.

0·1 mm

(b) Interpretation of pattern.

PLATE VII

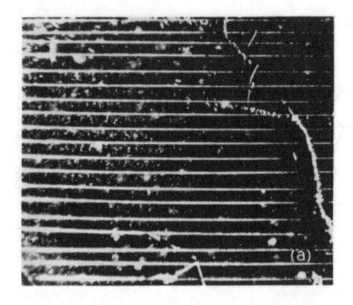

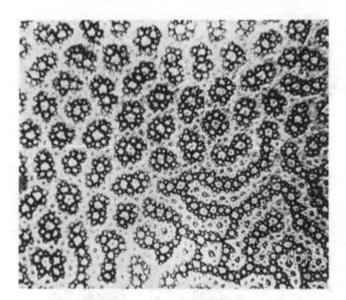

Patterns obtained on a cobalt crystal.
(*a*) parallel, (*b*) perpendicular, to hexagonal axis.

CHAPTER V

DOMAIN WALLS

5.1. Approximate treatment of wall width and energy

Before discussing conditions in the transition layer between two domains in more detail we can make a rough estimate of the width of the layer and of its energy per unit area. We write the 'wall energy', γ, as the sum of two terms. The first arises because the magnetization vector inside the wall is not in an easy direction as defined by the magnetocrystalline anisotropy; this term will be of order of magnitude $K\delta$ per unit area, where K is the anisotropy constant and δ the thickness of the wall. The second term arises because the direction of magnetization is changing inside the wall, so that the Heisenberg exchange energy is not a minimum; the exchange interactions are supposed to be important only between 'nearest neighbours', and we can therefore write the energy term as $A(1 - \cos \epsilon)\,\delta$ per unit area of wall, where A is the exchange energy per unit volume, a the interatomic distance and ϵ the angle between the spins of neighbouring atoms as we go through the wall (ϵ is here assumed to be small and constant throughout the wall). If the wall separates domains whose directions of magnetization differ by π, we can put $\epsilon = \pi a/\delta$ and so obtain for the total wall energy

$$\gamma = K\delta + Aa^2\pi^2/2\delta.$$

The value of δ will be that which makes γ a minimum, i.e.

$$\delta = \frac{\pi a}{\sqrt{2}} \sqrt{\frac{A}{K}},$$

and the corresponding value of γ is

$$\gamma = \sqrt{2}\,\pi a \sqrt{(AK)}.$$

The value of the exchange energy, A, is of the order of 2×10^{10} ergs/c.c. in iron, and K, the magnetocrystalline energy, is about 4×10^5 ergs/c.c. The atomic spacing, a, is 4×10^{-8} cm., so that we obtain $\delta \sim 2 \times 10^{-5}$ cm. and $\gamma \sim 10$ ergs/cm.[2]. In materials with low magnetocrystalline anisotropy and large stresses, the favourable directions of magnetization may be determined by magnetostriction effects, and the quantity K in the above calculation must be replaced by the magnetostrictive energy $C = \lambda \sigma$ ($\lambda =$ saturation magnetostriction coefficient, $\sigma =$ stress).

The values just obtained for δ and γ are, of course, only very rough approximations. It is clear that no very precise calculation can be made, because the exchange forces between atoms play an important part in deciding conditions inside the walls and our knowledge of exchange forces is not accurate or detailed. Estimates of the exchange energy per unit volume, for instance, vary widely according to the way in which the experimental results are treated (Stoner, 1950). Although we cannot hope to find the absolute value of the wall energy very precisely, it is nevertheless useful to carry out calculations rather more detailed than those made above in order to find out the way in which the wall energy might be expected to depend on various factors, such as the orientation of the wall and of the domain magnetizations and the magnitude of the crystal anisotropy, the magnetostriction coefficient and the stress.

5.2. Change of spin direction within a wall

In the last section we assumed that the direction of magnetization (the direction of the spin moments of individual atoms) changed at a constant rate, specified by the angle ϵ, throughout the wall. In a real wall the spin direction will change in some more complicated way. If we define the spin direction by angular coordinates, ϕ being the angle it makes with the normal to the wall and θ the angle with some fixed axis in the plane of the wall (see fig. 41), then θ and ϕ will be functions of position in the wall (functions of one coordinate—x, directed along the wall normal— if the wall is plane), and the problem is to choose the functions so as to make the energy of the wall a minimum.

This minimum problem was first treated and solved by Bloch (1932) with the assumption that θ remained constant throughout the wall. Becker and Döring give a similar treatment, but point out that the assumption $\theta =$ constant implies a divergence of magnetization within the wall which will create a local magnetic field and augment the wall energy. This point has been taken up by Néel (1944 b), who shows that the energy of the local field would be very large and suggests that a more reasonable assumption is ϕ = constant, which gives zero divergence of magnetization within the wall. It was shown in Chapter IV that the wall orientation would be such that the component of magnetization normal to the wall was the same on both sides of the wall. The constancy of ϕ within the wall means that the normal component remains the same even inside the wall. In any real wall the condition ϕ = constant will be

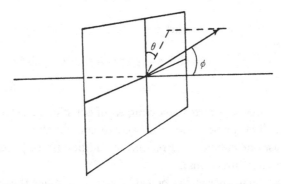

Fig. 41. Angular coordinates of spin direction within a wall.

modified somewhat in order to reduce the exchange and anisotropy energies, but the magnetostatic energy associated with changes in ϕ is so large that in most materials it is a good approximation to treat ϕ as exactly constant, as will be done throughout the following discussion.

5.3. Energy in walls

The first stage in calculating the wall thickness and energy is to express the two components of energy—magnetocrystalline and exchange—as functions of the polar coordinates θ and ϕ. The magnetocrystalline energy has to be obtained by expressing equation (2.1) in terms of θ and ϕ. The expressions are, in general, complicated, but in the cases where the normal to the wall (the polar axis, ϕ = 0) coincides with an axis of symmetry of the crystal they become simpler, and Néel gives the following results for the magnetocrystalline energy per unit volume, W_K:

(i) Wall normal a cube edge direction ([100] axis)

$$W_K = K_1[\sin^2 \phi - \tfrac{7}{8} \sin^4 \phi - \tfrac{1}{8} \sin^4 \phi \cos 4\theta]. \tag{5.1}$$

(ii) Wall normal a cube diagonal ([111] axis)

$$W_K = K_1\left[\frac{\cos^4 \phi}{3} + \frac{\sin^4 \phi}{4} - \frac{\sqrt{2}}{3} \cos \phi \sin^3 \phi \cos 3\theta\right]. \tag{5.2}$$

(iii) Wall normal a cube face diagonal ([110] axis)

$$W_K = \frac{K_1}{4}\left[1 - 4 \sin^2 \phi + 4 \sin^4 \phi \right.$$
$$\left. + (6 \sin^2 \phi - 4 \sin^4 \phi) \sin^2 \theta - 3 \sin^4 \phi \sin^4 \theta \right]. \tag{5.3}$$

In these expressions terms involving K_2 of equation (2.1) have been neglected. The plane $\theta = 0$ is chosen to contain the wall normal and at least one cube edge direction in addition (in (iii) it contains two cube edge directions).

The exchange energy can be calculated if we make the assumption that exchange forces act only between nearest-neighbour atoms. (This assumption is probably untrue; see, for example, Stoner, 1948.) If we write the total exchange energy per unit volume, with all spins parallel, as A, then the contribution from each pair of atoms is A/a^3n, where a is the interatomic spacing and n the number of 'nearest neighbours' to any given atom. If two spins have an angle ϵ between them, the increase in exchange energy is $\dfrac{A}{a^3n}(1 - \cos \epsilon)$, which for small ϵ is equal to $\dfrac{A\epsilon^2}{2a^3n}$. In terms of the coordinates θ and ϕ we can write $\epsilon = \sin \phi \, \delta\theta$, since ϕ is constant throughout the wall; $\delta\theta$ is the change in θ in going from one atom to the neighbour in question. If we take the axis of x along the normal to the wall and write λ for the angle between the normal and the line joining the neighbouring atoms, we have

$$\delta\theta = \frac{d\theta}{dx}\, \delta x = \frac{d\theta}{dx}\, a \cos \lambda,$$

and so obtain for the increase in exchange energy of the pair of atoms in question

$$W_E = \frac{A \sin^2 \phi}{2an} \left(\frac{d\theta}{dx}\right)^2 \cos^2 \lambda.$$

The increase in exchange energy between a given atom and all its neighbours is obtained by multiplying W_E by the number of neighbours, n, and replacing $\cos^2 \lambda$ by its average value for all the neighbours. This latter average is $\frac{1}{3}$ for all types of cubic symmetry. The total number of pairs of atoms in unit volume is a^3, so that we obtain finally, for the increase in exchange energy per unit volume,

$$W_E = \frac{A}{6} a^2 \sin^2 \phi \left(\frac{d\theta}{dx}\right)^2.$$

The total excess energy per unit area of the wall, referred to the energy in the surrounding domains as zero, can be written

$$\gamma = \int_{-\infty}^{+\infty} (W_K - W_{K0} + W_E)\, dx, \tag{5.4}$$

where W_{K0} is the magnetocrystalline energy density in the surrounding domains. If we put

$$W_K - W_{K0} = f(\theta) \tag{5.5}$$

and

$$\frac{A}{6} a^2 \sin^2 \phi = E,$$

this becomes

$$\gamma = \int_{-\infty}^{+\infty} \left[E\left(\frac{d\theta}{dx}\right)^2 + f(\theta) \right] dx. \tag{5.6}$$

The actual variation of θ with x in the wall will be that required to make γ a minimum. Following Néel and Becker, we can solve this variation problem by considering the effect of a small arbitrary change, $\delta\theta$, in θ. We have

$$\delta\gamma = \int_{-\infty}^{+\infty} \left[2E\left(\frac{d\theta}{dx}\right) \frac{d}{dx}(\delta\theta) + \frac{df(\theta)}{dx} \delta\theta \right] dx.$$

The first term can be simplified by using the relation

$$\frac{d\theta}{dx} \frac{d}{dx}(\delta\theta) = \frac{d}{dx}\left(\frac{d\theta}{dx} \delta\theta\right) - \delta\theta \frac{d^2\theta}{dx^2},$$

s

and the fact that the variation being considered must be one with $\delta\theta = 0$ at each of the limits of integration. The term $\dfrac{d}{dx}\left(\dfrac{d\theta}{dx}\,\delta\theta\right)$ therefore vanishes on integration and we are left with

$$\delta\gamma = \int_{-\infty}^{+\infty}\left[\frac{df(\theta)}{dx} - 2E\,\frac{d^2\theta}{dx^2}\right]\delta\theta\,dx.$$

For γ to be a minimum, $\delta\gamma$ must be zero whatever the form of $\delta\theta$; the quantity in square brackets must therefore be zero, and we obtain

$$\frac{df(\theta)}{dx} = 2E\,\frac{d^2\theta}{dx^2}$$

or

$$f(\theta) = E\left(\frac{d\theta}{dx}\right)^2 + C.$$

Since, in the domains on each side of the wall, we have the limiting conditions $f(\theta) = 0$, $d\theta/dx = 0$, we can put $C = 0$ and get the simple result

$$f(\theta) = E\left(\frac{d\theta}{dx}\right)^2. \tag{5.7}$$

This shows that at every point in the wall the increase in magneto-crystalline energy is just equal to the increase in exchange energy. We can therefore write the total wall energy as twice its magneto-crystalline part

$$\gamma = 2\int_{-\infty}^{+\infty} f(\theta)\,dx,$$

or, taking account of equation (5.7),

$$\gamma = 2\sqrt{E}\int_{\theta_1}^{\theta_2}\sqrt{\{f(\theta)\}}\,d\theta$$

$$= a\sin\phi\sqrt{(2A/3)}\int_{\theta_1}^{\theta_2}\sqrt{\{f(\theta)\}}\,d\theta, \tag{5.8}$$

where θ_1 and θ_2 are the values of θ in the domains on either side of the wall.

The wall energy can therefore be calculated in any case for which the angles θ_1, θ_2 and ϕ and the magnetocrystalline energy function $f(\theta)$ are known. In iron in low magnetic fields the domains will all be magnetized along cube edge directions, so that we have to do with '90°' and '180°' walls only. The magnetocrystalline energy has already been given in terms of θ and ϕ for three simple orientations of wall (p. 96). For 90° walls with these orientations we have also:

(i) $\phi = 90°$, $\theta_1 = 0°$, $\theta_2 = 90°$;

(ii) $\cos \phi = 1/\sqrt{3}$, $\theta_1 = 0°$, $\theta_2 = 120°$;

(iii) $\phi = 45°$, $\theta_1 = 0°$, $\theta_2 = 180°$;

and in all cases $W_{K0} = 0$, since the magnetization in the domains lies in easy directions. Using these values in equations (5.1), (5.2), (5.3), (5.5) and (5.8), we get the following results for the energy of 90° walls:

$$\text{(i)} \quad \gamma_{100} = \frac{a}{2} \sqrt{(\tfrac{2}{3}AK_1)}, \tag{5.9}$$

$$\text{(ii)} \quad \gamma_{111} = \frac{32}{27} \frac{a}{2} \sqrt{(\tfrac{2}{3}AK_1)}, \tag{5.10}$$

$$\text{(iii)} \quad \gamma_{110} = 1 \cdot 727 \frac{a}{2} \sqrt{(\tfrac{2}{3}AK_1)}. \tag{5.11}$$

It was shown in the last chapter that the orientation of 90° walls was controlled by demagnetizing and magnetostriction effects. The walls actually occurring will not, in fact, be the ones with minimum surface energy, those with normals along cube edge directions, but rather those of types (ii) or (iii) with somewhat larger energies.

180° walls can be regarded as pairs of 90° walls close together, and, since neither demagnetizing fields nor magnetostriction impose any constraint on their orientation, the most favourable orientation will be that of type (i), because this type has the lowest surface energy; we have therefore

$$\gamma_{180°} = a\sqrt{(\tfrac{2}{3}AK_1)}. \tag{5.12}$$

When a moderate magnetic field ($\sim$ 100 gauss) is applied to iron the domain magnetizations all turn to the easy directions nearest to the field and then rotate from the easy directions towards the field, as was described in Chapter II. There are therefore no longer any 180° walls, and instead of 90° walls there are walls between domains whose magnetization directions differ by less than 90°. The wall energy can be calculated just as before, but the value of ϕ will depend on the field strength (§ 2.2), and the value of W_{K0} will no longer be zero but must be found by putting

$\theta = 0$ and the appropriate value of ϕ in equation (5.1), (5.2) or (5.3). There will be an increase in magnetic potential energy of the wall, because of the action of the magnetic field on its spins, but, because ϕ is constant inside the wall and on both sides of it, this increase will be just the same as in the material surrounding the wall and the excess energy of the wall will not be affected.

We can illustrate the calculation of wall energy in a finite magnetic field by reference to the particular case of a long single crystal magnetized in a [110] direction, as considered in the preceding chapter. As shown in fig. 33, the walls run across the crystal, perpendicular to the [110] direction, so that they are of type (iii). W_K can therefore be found from equation (5.3); θ changes from 0 on one side of a wall to π on the other, so that W_{K0} is given by the value of equation (5.3) with $\theta = 0$ and the wall energy is found by integrating equation (5.8) from 0 to π. The final result, as given by Néel, is

$$\gamma = \frac{a}{2} \sqrt{(\tfrac{2}{3}AK_1)} \sin \phi \left[\sin \phi \sqrt{(6 - 4 \sin^2 \phi)} + \frac{6 - 7 \sin^2 \phi}{\sqrt{3}} \sin^{-1} \sqrt{\frac{3 \sin^2 \phi}{6 - 7 \sin^2 \phi}} \right],$$

$$(5.13)$$

where ϕ, the angle between domain magnetizations and the field, is given by the equation

$$H = \frac{2K_1}{I_s} \cos \phi (2 \cos^2 \phi - 1).$$

The variation of γ with H according to these equations has already been quoted in Table I (p. 79).

5.4. The width of walls

In order to find the way in which the spin direction changes as we traverse a wall, we must integrate equation (5.7), which can be rewritten

$$dx = \sqrt{E} \frac{d\theta}{\sqrt{\{f(\theta)\}}}. \qquad (5.14)$$

$f(\theta)$ is usually too complicated to make integration easy, but in the case of a 90° wall perpendicular to a cube edge direction, equations (5.1) and (5.5), together with the condition $\phi = \tfrac{1}{2}\pi$, give

$$f(\theta) = K \sin^2 \theta \cos^2 \theta,$$

so that equation (5.14) can be integrated, giving

$$x = \sqrt{\left(\frac{E}{K}\right)} \log \tan \theta. \qquad (5.15)$$

As this equation shows, θ approaches its limiting values in the domain on either side of the wall only asymptotically, but it is reasonable to say that the 'width of the wall' is given by $3\sqrt{(E/K)}$, since 75% of the change in θ is completed in this distance. If a 180° wall is considered as made up from two 90° walls, the difficulty arises that the distance between the two 90° parts would be infinite. Néel shows that this difficulty is removed by allowing for a magnetostrictive term; the material between the two 90° sections of the 180° wall is in a state of strain, and energy considerations therefore require it to be as thin a layer as possible. The term involved is a small one and has no appreciable effect on the energy of the wall already calculated; its only effect is to replace the asymptotic variation of θ for large x values by a more abrupt one, and thus to make the total width of the 180° wall finite and roughly equal to the width of two 90° walls.

The calculations of wall energy and thickness have been carried through for a cubic material with easy directions along the cube edges, such as iron. Similar calculations can, of course, be made for materials with different magnetocrystalline properties, such as nickel and cobalt. Lilley (1950) has recently outlined the general treatment of wall properties and derived results for all the special cases of interest.

5.5. Walls in materials under stress

We have so far only considered walls in materials where the magnetocrystalline anisotropy is the chief factor determining direction of magnetization. The calculations can, however, be applied quite easily to the case where a large stress acts and the magnetostrictive energy $\lambda\sigma$ is much more important than the magnetocrystalline energy K. The favourable direction of magnetization will then be determined chiefly by the stress, and the walls between domains will be regions of balance between exchange energy and magnetostrictive energy. The calculation of wall energy is precisely the same as that already carried out, if $f(\theta)$

in equation (5.5) is taken to be the excess magnetostrictive energy, rather than the excess magnetocrystalline energy. If the stress is tensile for a material with positive magnetostriction or compressive for one with negative magnetostriction, there will be only two favourable directions of magnetization and the walls will be 180° ones. The magnetostrictive energy is given by equation (3.20), so that we have $f(\theta) = \frac{3}{2}\lambda\sigma\cos^2\theta$, if the magnetostriction is isotropic. Solution of equation (5.8) with this value of $f(\theta)$ gives the wall energy

$$\gamma = 2a\sqrt{(A\lambda\sigma)}. \qquad (5.16)$$

It is, of course, only in rather special materials that either the magnetocrystalline or the magnetostrictive forces are so large as to justify complete neglect of one of them in comparison with the other. If both have to be taken into account, exact calculation of wall energies becomes complicated (see for example, Lilley, 1950), but in any case stress and orientation conditions are seldom known in sufficient detail for more than rough averages to be calculated. The variations of wall energy produced by local variations of stress are a very important factor controlling the movements of walls; this matter will be considered further in the next chapter.

5.6. Numerical estimates of wall energy and thickness

The surface energy, γ, and thickness, δ, of a wall are given by equations (5.8), (5.15) and (5.16), in terms of magnetocrystalline and magnetostriction constants, K and λ, and the exchange energy per unit volume, A. K and λ are known fairly precisely for many materials (see Chapters II and III), but estimates of A vary widely. A was defined as the exchange energy per unit volume and is thus the same as the quantity E used by Néel (1944b) or $\frac{1}{2}n$ times the quantity I of Becker and Döring (1939, p. 192). In terms of the Weiss theory, we can write

$$A = \frac{1}{2}NI_s^2,$$

N being the 'molecular field coefficient'. The simplest way of finding A is from the experimental value of the Curie temperature, using the equation of the simple Weiss theory,

$$\theta = \frac{NI_s^2 M}{3R\rho q},$$

where q = effective number of spins per atom, R = gas constant, M = molecular weight, and ρ = density. For iron we have

$$\theta = 1040^\circ \text{ K.}, \quad I_s = 1720 \text{ gauss}, \quad M = 55\cdot8, \quad \rho = 7\cdot8, \quad q = 2\cdot2,$$

and so obtain $\qquad A = 2\cdot6 \times 10^{10} \text{ ergs/c.c.}$

With this value of A, the energy of a 180° wall in iron, as given by equation (5.12), becomes

$$\gamma = 3\cdot0 \text{ ergs/cm.}^2,$$

and the values for 90° walls with various orientations, as given by equations (5·9), (5.10) and (5.11), are of the same order of magnitude. Néel's estimates of wall energy differ by a factor of about 2 from these values, because he uses the experimental Curie constant of iron above its Curie point to find the molecular field coefficient N. Other estimates of the exchange energy can be made from observations of magnetization at very low temperatures (Lifshitz, 1944). The variety of values of A that can be derived by these different methods reflects the inadequacy of present-day theory and emphasizes that theoretical estimates of wall energy are unlikely to be at all accurate. More detailed discussions of the most probable values of wall energies are given in the review articles by Kittel (1949a) and Stoner (1950).

5.7. Experimental measurement of wall energy

Wall energies calculated by the methods just described were used in the discussion of domain arrangements in the preceding chapter and will be used again in the next chapter in considering magnetization by movement of walls. In both cases experiment supports the order of magnitude of wall energies deduced theoretically. In order to make any more precise comparison between theory and experiment it is necessary to study cases where the arrangement of walls is known in detail and is reasonably simple. Two such cases are known. The first is the single-crystal 'picture-frame' specimen of Williams and Shockley which was described on p. 86, with essentially a single 180° wall running right round the specimen. It may well prove possible to obtain much information as to the properties of walls from such a specimen, but so far few measurements have been reported. A possible method of investigating the wall thickness and energy by means of a neutron beam has been suggested by Newton and Kittel (1948).

The other simple case has been known for a longer time and extensively studied by Sixtus and Tonks (1931, 1932, 1933), Preisach (1932) and Döring and Haake (1938). It is the case of a material with positive magnetostriction and small magneto-crystalline anisotropy, when it is placed under tension. In such material there are only two favourable directions of magnetization, antiparallel to one another and lying along the line of tension.

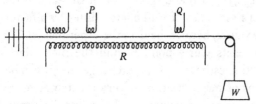

Fig. 42. Schematic arrangement for the study of magnetic changes in a wire under tension.

Magnetization curves with various amounts of tension were shown in fig. 22; when the applied tension is sufficient to overcome local irregularities caused by internal stresses and crystal anisotropy the specimen is seen to be always magnetized to saturation in one direction or the other, reversal of the direction of magnetization taking place discontinuously at critical values of the applied field.

Sixtus and Tonks studied the mechanism of this reversal with the apparatus shown schematically in fig. 42. They found that if a large reverse field was applied locally by the coil S, the main field provided by R being too small to cause reversal, a small region of reversed magnetization was formed inside S and spread rapidly along the specimen until the magnetization of the whole had changed direction. The coils P and Q were used to measure the speed with which the reversal spread; the time interval between voltages induced in P and Q measured the longitudinal speed, while the form of pulse induced in P or Q (observed by oscillograph) indicated how the reversal spread across the specimen.

Assuming circular symmetry in their wire specimens, Sixtus and Tonks deduced that the reversal was propagated as shown in fig. 43. A single curved 180° wall separated the two parts of the wire magnetized in opposite directions and moved along the wire as the reversal proceeded. The shape of the wall and its speed of propagation depend on eddy currents, demagnetizing fields and wall energy, and can give useful information about the properties of walls in motion, as will be shown in Chapter VII.

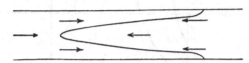

Fig. 43. Approximate form of a boundary moving from left to right. Arrows show directions of magnetization.

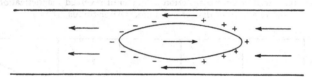

Fig. 44. Sketch showing a nucleus of reversed magnetization within a wire.

Information about the wall energy can, however, be obtained more readily from a study of the early stages of the reversal, as was shown by Döring (1938). Sixtus (1935) showed that if the 'starting field', H_s, provided by coil S was applied for a short time only, then the reversal of magnetization did not spread down the wire, although a small nucleus of reversed magnetization was formed near S and remained there, as shown in fig. 44, even when the field H_s was cut off. These 'frozen-in' nuclei could be detected, and their shape determined, by sliding a search coil connected to a ballistic galvanometer along the wire. Fig. 45 shows the way in which the cross-section of three of Sixtus's nuclei varied along the wire. The volume of the largest is about 5×10^{-4} c.c., so that we have here a single domain of quite large size whose shape is known. If the general field H caused by coil R in fig. 42 is increased, a critical field can be reached at which the nucleus 'bursts' and produces a wall travelling down the wire, as in fig. 43.

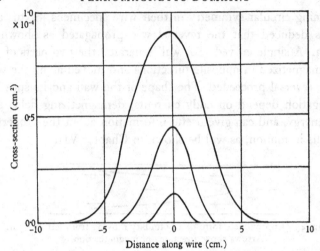

Fig. 45. Variation of cross-section of nuclei of reversed magnetization along a wire of diameter 0·3 cm. (After Sixtus, 1935.)

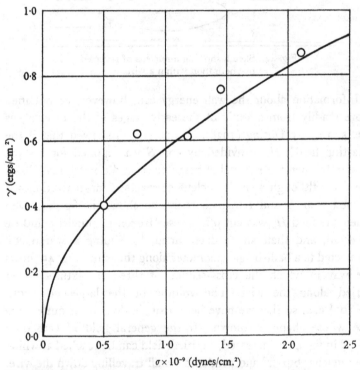

Fig. 46. Variation of wall energy, γ, with tension, σ, in a Permalloy wire. Theoretical curve from equation (5·16), experimental points form Döring and Haake (1938). (After Becker and Döring, 1939.)

The study of this critical field in relation to the length and thickness of the nucleus enabled Döring to calculate values for the surface energy in the wall surrounding the nucleus. In a simple case the nucleus of fig. 44 is prevented from growing only by the fact that a larger nucleus would have a larger surface area and therefore a larger wall energy; the nucleus will grow if this increase in wall energy is less than the decrease in magnetic energy, $2HI_s dV$, due to the change in volume dV, otherwise it will shrink. The problem is, however, complicated by the energy of the demagnetizing field of the nucleus and by the fact that there is a 'frictional' resistance to the movement of the wall which opposes either growth or shrinkage. A more complete calculation is given by Becker and Döring (1939). Results obtained by Döring and Haake (1938) for the wall energy, γ, as a function of σ, the tension applied to the wire, are shown in fig. 46. Equation (5.16) gives the wall energy as $\gamma = 2a\sqrt{(A\lambda\sigma)}$, and the curve drawn in fig. 46 shows this theoretical value, assuming $a = 3\cdot6 \times 10^{-8}$ cm., $\lambda = 2 \times 10^{-5}$, $A = 3\cdot0 \times 10^9$ ergs/c.c. Similar results have been obtained more recently by Ogawa (1949). The agreement is considerably better than would be expected from the approximate way in which exchange forces have been treated in deriving equation (5.16) for γ.

HINDRANCES TO DOMAIN WALL MOVEMENTS

6.1. Introduction

In Chapters II and III we considered the way in which the energy of a ferromagnetic domain depended on the direction of its magnetization relative to the crystal structure and to the mechanical stresses. In many cases we were able to deduce the way in which the equilibrium arrangement of domains changed when a magnetic field was applied and so to obtain magnetization curves. Many of these curves indicate an abrupt change in magnetization when the field is changed from an infinitesimal positive to an infinitesimal negative value. This is to be expected, because in an ideally homogeneous material the magnetocrystalline and magnetostrictive energies define several exactly equivalent easy directions and, in the absence of a field, there is nothing to decide the distribution of domain magnetizations among these directions and so determine the overall magnetization. Experiments do not show any such large and abrupt change in magnetization—indeed, most of the phenomena of technical interest in ferromagnetics are concerned with the finite slopes of magnetization curves in regions where, according to the equations of Chapters II and III, the slope should be infinitely steep.

In order to explain these phenomena we must abandon the hypothesis of ideally homogeneous materials that has been used so far and take account of the small accidental local variations of stress and composition that occur in all real materials. Such irregularities can provide additional energy terms in various ways and so distinguish between domain arrangements which would be exactly equivalent in a homogeneous material. While it is possible to make rough estimates of the size of these terms and of their effects, it is unfortunate that just in the region where technical interest is greatest it has so far proved impossible to make precise quantitative calculations of any widespread application. There are two types of difficulty, the first in specifying the additional energy terms and their relation to inhomogeneities in the material, the second in finding the effect of several such terms acting in combination. In a few specially simple cases the difficulties are lessened and fairly precise calculations of properties can be made; but for the majority of materials theory cannot give more than a semi-quantitative guide to their behaviour in low fields. We shall begin by outlining the ways in which it has been suggested that inhomogeneities in the material can affect magnetic processes, taking the simplest possible models of domains and their boundaries. The ways in which observable magnetic properties depend on these effects in real materials will be considered later.

6.2. The effect of internal stresses on domain volume energy

The simplest effect of random internal stresses has been considered by Becker (1932). If we have a material with large magnetocrystalline anisotropy, there will be, in zero field, a number of equivalent easy directions determined by the anisotropy (six in iron, eight in nickel). If a small stress acts, it will make one pair of the easy directions more favourable than the others, owing to the additional magnetostrictive energy. Stresses varying in direction and magnitude from point to point will thus cause the most favourable direction to vary from point to point, remaining always very close to one or other of the crystal easy directions, so long as the stress energy is much smaller than the magnetocrystalline component. The equilibrium arrangement of domains in zero field will be one where every domain has its magnetization in the local 'most favourable' direction determined in this way by the stresses. A finite field will be required to change the magnetization from this equilibrium state. We can calculate the relation between field and magnetization for a simple model (fig. 47).

We suppose that the stress, σ_i, varies sinusoidally between tension and compression along the x-axis, as shown in fig. $47a$. If the material has six mutually perpendicular easy directions, as has iron, a zero-field arrangement of domains might be as shown schematically in fig. $47b$. Application of a field H along the x-axis will cause domains (i), (ii) and (iv) to grow at the expense of (ii) and (iii), as shown in fig. $47c$, the movement of the walls continuing until the decrease in magnetic energy, HI_s, is balanced by

the increase in magnetostrictive energy $\lambda \sigma_i \sin ax$. For small movements we can write
$$HI_s = \lambda \sigma_i ax,$$
or, since the change in magnetization in the x-direction is $I_s x$,

$$\Delta I = \frac{I_s^2}{a \lambda \sigma_i} H, \tag{6.1}$$

where ΔI is the contribution to the magnetization from the movement of unit area of one wall. We shall see later that the calculation can be generalized to give averages over all the walls and possible orientations of walls in a material.

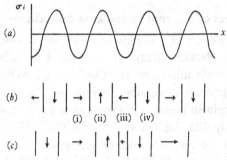

Fig. 47. Model illustrating movements of domain walls in a material with fluctuating internal stresses, when a field is applied. (a) Variation of tensile stress along x-axis. (b) Corresponding arrangement of domains in zero field (schematic). (c) Domain arrangement when a small field acts from left to right.

It is clear that this volume-energy effect of stresses cannot influence the position or movement of 180° walls, the walls between domains with antiparallel magnetizations, since the magnetostrictive energy is unaffected by a reversal of the direction of magnetization. It can, however, be shown (Kondorsky, 1937; Kersten, 1938) that there is another effect of stresses which hinders the free movement of 180° walls and so prevents the appearance of infinite permeabilities. This is the effect of stresses on the surface energy of walls.

6.3. The effect of internal stresses on domain wall energy

The surface energy, γ, of the wall between two domains has been calculated in Chapter V for the two extreme cases where magnetocrystalline (equation (5.16)) or magnetostrictive (equation

(5.12)) effects could be ignored. If varying internal stresses are present it is clear that γ will be modified, either by the addition of a magnetostrictive term to equation (5.12) or by the actual variation of σ in equation (5.16). This latter case, where γ is primarily determined by some uniform stress, but is locally modified by stress fluctuations, has been considered in some detail by Kersten (1938). He deals with two cases, illustrated in fig. 48.

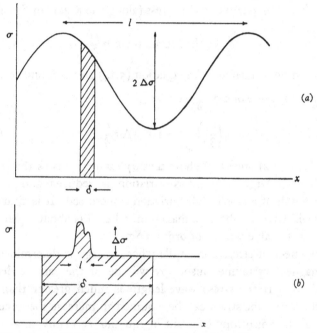

Fig. 48. (a) Wall thin compared with stress fluctuation wave-length. (b) Wall thick compared with stress fluctuation wave-length.

In fig. 48a, the width of the wall, δ, is small compared with the wave-length of the stress, l; in fig. 48b, δ is large compared with l. The quantity of interest is $\partial\gamma/\partial x$, the rate of change of γ with position, due to the presence of the stress variation, $\Delta\sigma$. In fig. 48a we can regard σ as constant throughout the wall and hence obtain, by differentiating equation (5.16),

$$\frac{\partial\gamma}{\partial x} = \tfrac{3}{2}\lambda\delta\,\frac{\partial\sigma}{\partial x},$$

and since $\dfrac{\partial \sigma}{\partial x} \approx \dfrac{2\Delta\sigma}{l}$, we have

$$\frac{\partial \gamma}{\partial x} = 3\lambda\Delta\sigma\frac{\delta}{l}. \qquad (6.2)$$

In fig. 48b we can say that the stress variation adds an extra magnetostrictive energy $\frac{3}{2}\lambda(\frac{1}{2}l\Delta\sigma)\sin^2\theta$ to the normal wall energy, where θ is the angle between the stress direction and the direction of magnetization within the wall. θ, of course, varies as the wall moves relative to the stress (along the x-axis in fig. 48b). We can write

$$\frac{\partial \gamma}{\partial x} = \frac{3}{4}\lambda l\Delta\sigma . 2\sin\theta\cos\theta\,\frac{\partial\theta}{\partial x}.$$

$\partial\theta/\partial x$ can be evaluated from equation (5.14) and is found to have a maximum value of $0\cdot77\dfrac{1}{\delta}$, so that

$$\left(\frac{\partial \gamma}{\partial x}\right)_{\text{max.}} = 1\cdot15\lambda\Delta\sigma\frac{l}{\delta}. \qquad (6.3)$$

Equations (6.2) and (6.3) show that $\partial\gamma/\partial x$ increases as the wave-length of a large-scale stress variation is reduced and as the wave-length of a small-scale variation is increased. It is clear that $\partial\gamma/\partial x$ will have an absolute maximum when δ is about equal to l, and that its value will be of order $\lambda\Delta\sigma$.

The effect of stresses on walls whose energy is determined by the magnetocrystalline anisotropy will be of the same order of magnitude. If the stress wave-length is much greater than the wall thickness, the stress can be assumed to increase the effective value of the anisotropy constant in equation (5.12) by an amount of order $\lambda\Delta\sigma$; if the wave-length is much smaller than the wall thickness, the stress adds to the wall energy just as in the derivation of equation (6.3). In both cases the size of $\partial\gamma/\partial x$ is clearly of order $\lambda\Delta\sigma$. There is little point in attempting to make more precise estimates of these wall-energy effects, because they depend so much on the details of the internal stress variations, and these are not known, in practice.

6.4. Effects of 'inclusions'

In many materials the effects of varying internal stresses can account for the observed magnetic properties in low fields. In others, particularly those with low magnetostriction, these effects are less important than those caused by variations in the composition of the material, first considered in detail by Kersten (1943). For simplicity we may consider the case illustrated in fig. 49, which shows a small spherical region of non-magnetic material embedded in the ferromagnetic. Such an inclusion might consist of a bubble of gas, for example, or, in iron, a deposit of cementite.

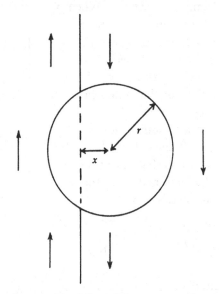

Fig. 49. Intersection of the wall between two domains with a spherical non-magnetic inclusion.

If the boundary wall between two domains cuts the inclusion, as shown in fig. 49, it is clear that the total wall area in the ferromagnetic will be a minimum when the wall bisects the inclusion. If the wall is displaced, its area, and consequently its surface energy, will increase. A finite magnetic field will thus be required to displace the wall from its symmetrical position through the inclusion. The change in position caused by a field H acting in the direction of the magnetization can be found from the conditions for minimum energy. The change in magnetic energy for a wall of unit area when the wall moves a distance x is $2HI_s x$. The change in wall area is given by $\pi(r^2 - x^2)$ for each inclusion; if we suppose that there are n inclusions per unit volume and that they are distributed regularly at the corners of a cubic lattice, the change in wall energy becomes $n^{\frac{2}{3}}\pi(r^2-x^2)\gamma$ for a wall of unit area. To make the sum of these two energy terms a minimum we must have

$$2HI_s = 2\pi n^{\frac{2}{3}}\gamma x. \tag{6.4}$$

The wall displacement, and hence the change in magnetization, is thus proportional to the field H so long as $x < r$, the volume susceptibility being $\chi = 2I_s^2/\pi n^{\frac{2}{3}}\gamma$. When $x = r$ and the wall just touches the edge of the inclusion, there is no longer any force opposing the movement, and the wall moves on to the next set of inclusions (fig. 50). If all the inclusions are of the same size, the field required to force the wall through one set will suffice also to move it through all the others. The field making $x = r$ is therefore a critical field for our idealized material, and can be called its coercive force. It is given, according to equation (6.4), by

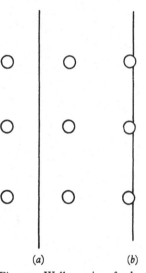

$$H_c = \frac{\pi n^{\frac{2}{3}}\gamma r}{I_s}. \tag{6.5}$$

(a)   (b)

Fig. 50. Wall moving freely between inclusions (a) and restrained by intersection with inclusions (b).

For iron containing 1 % of impurities in the form of spherical inclusions 1μ in diameter the value of H_c is approximately 1 oersted. Kersten made more detailed calculations, considering the way in which H_c and χ depended on the shape and total volume of the inclusions and on the temperature. He showed that in many cases the behaviour of real materials corresponded closely to that of his inclusion model.

Kersten's whole treatment of the effects of inclusions has, however, been criticized by Néel (1946). He pointed out that Kersten had neglected the effects of demagnetizing fields in the material around the inclusions, and that these fields can add an energy far greater than the simple wall surface energy included in Kersten's theory. Fig. 51 illustrates two cases. In the first, a wall bisects an inclusion; in the second, the wall has moved to one side and the inclusion is entirely within one of the domains. The north and south poles produced on the boundaries of the inclusion are shown by plus and minus signs. In fig. 51b the energy of the demagnetizing field around the spherical inclusion can be written as $\frac{1}{2}NI_s^2=\frac{8}{9}\pi^2r^3I_s^2$, since the demagnetizing coefficient N of a sphere is $\frac{4}{3}\pi$. In fig. 51a the energy will clearly be less, and the difference in energy between the two cases will be of order $4r^3I_s^2$. The ratio of the change in demagnetizing energy

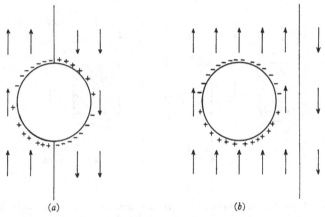

(a)   (b)

Fig. 51. Free poles induced on the boundary of a non-magnetic inclusion.
(a) With wall intersecting it. (b) With no intersecting wall.

to the change in wall energy on going from fig. 51a to fig. 51b is therefore $\dfrac{4r^3I_s^2}{\pi r^2\gamma}\approx\dfrac{rI_s^2}{\gamma}$. Since I_s is 1720 gauss and γ about 1 erg/cm.^{2} for iron, the demagnetizing energy is more important than the wall energy for all inclusions greater than about $\frac{1}{100}\mu$ in diameter, and Kersten's treatment, which entirely neglects the first energy term, is incorrect.

Néel suggests that arrangements such as those of fig. 51 are impossible because the magnetization in the immediate neighbourhood of the inclusion would be deflected by the large demagnetizing fields. In order to reach a stable state we must introduce small 'closing domains' similar to those found at the external boundaries of specimens in Chapter IV. When the wall bisects, or passes near, the inclusion it is possible to find an arrangement with no free poles;

an example for the simple case of a cubic inclusion is shown in fig. 52(a). If the inclusion is far from any wall, so that the total flux intercepted by it is not zero, then the appearance of free poles is inevitable. It does not follow, however, that the simple arrangement of fig. 52*b* is the one of minimum energy; as Néel points out, the energy of the demagnetizing field is greatly reduced by arrangements such as those in fig. 52*c*, where the free poles are spread out over a larger surface. The presence of these complicated

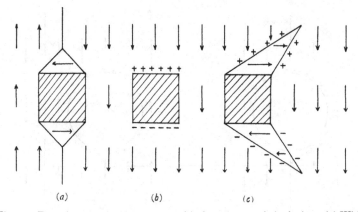

Fig. 52. Domain arrangements near a cubical non-magnetic inclusion. (*a*) With domain boundary; free poles eliminated by closing domains. (*b*), (*c*) Inclusions wholly within one domain; in (*c*) the density of free poles is reduced by the introduction of supplementary domains.

arrangements near each inclusion—the complexity will, of course, be greater for non-cubic inclusions—makes detailed calculation of the effect of the inclusion on wall movement impossible. Néel shows, however, that an upper limit to the coercive force can be deduced for the arrangement shown in fig. 52*c*, and considers that any other possible arrangement will give results of the same order. If a field is applied upwards in fig. 52*c*, the pointed supplementary domains will tend to grow longer and wider. Their size is determined, just as that of the nuclei in Sixtus's and Döring's experiments (§ 5·7), by the balance between wall energy, energy of the demagnetizing field, and energy in the applied field. As the applied field is increased, the supplementary domains will grow steadily until a critical field is reached at which unlimited growth is possible. This field, which can be identified with the coercive force, can be calculated roughly on the assumption that the supplementary domains have grown so long that the demagnetizing field energy can be neglected in comparison with the wall energy. Then, for unlimited growth, the change in magnetic energy, $HI_s dV$, for an increase in volume, dV, of the supplementary domains must exceed the change in wall energy, γdS, where dS is the corresponding increase in wall area. Assuming the supplementary domains to grow by the addition of cylindrical portions of radius r (the radius of the inclusion), we obtain for the critical field

$$H_c \approx \frac{\gamma}{I_s r}.$$

More exact calculation gives

$$H_c = \frac{5}{8} \sqrt{\left(\frac{\pi^3}{8}\right)} \frac{\gamma}{I_s r}. \tag{6.6}$$

This equation suggests that the coercive force should increase indefinitely as the size of the inclusions is reduced. When the inclusions are very small, however, the thickness of the domain walls becomes comparable with the diameter of the inclusions, and the arrangements shown in fig. 52 will clearly not occur. Calculations are not possible when the inclusions are of about the same dimensions as the wall thickness, but for inclusions small compared with the thickness, calculations of their effects can again be made.

6.5. Effects of larger amounts of non-ferromagnetics

The 'inclusions' considered in the last section were treated as small imperfections within, or at the boundary of, individual domains. If, however, the proportion of non-ferromagnetic material is high, we must consider the material as consisting of ferromagnetic particles embedded in a non-ferromagnetic medium. The behaviour of such materials has been investigated by Stoner and Wohlfarth (1948) and by Néel (1947). If the ferromagnetic particles are large enough to contain many domains each, they will have the normal properties of the ferromagnetic; the properties of the composite material will depend on the size, shape and spacing of the particles, but can be calculated without any fundamental difficulties (see, for example, Richards et al. 1950).

If, on the other hand, the particles are very small they may each consist of a single domain, and the properties of such particles may differ considerably from those of the bulk ferromagnetic.

We can estimate fairly easily the order of size below which it is energetically more favourable for a particle to remain a single domain rather than to be split up into more than one. A sphere of radius r uniformly magnetized to an intensity I_s has energy $\frac{1}{2} \cdot \frac{4}{3}\pi I_s^2 \cdot \frac{4}{3}\pi r^3$ in its demagnetizing field. This energy can be reduced if the sphere is split into domains magnetized in different directions,

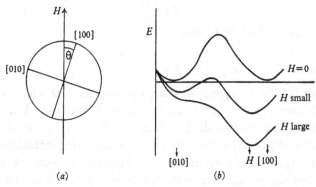

Fig. 53. (a) Field direction and crystal axes. (b) Angular variation of energy, E, in (100) plane for different values of field, H.

but at the cost of introducing wall energy, equal to $\pi r^2 \gamma$ for a diametral wall. We can thus say that the division into domains will be unprofitable if r is less than about γ/I_s^2, or, for iron, 10^{-6} cm. More exact calculations of the maximum sizes of 'single-domain' particles of various shapes have been made by Kittel (1946).

The only way in which a single-domain particle can change its magnetization is by a rotation of the magnetization of the whole particle; the processes of wall movement which are so important in normal ferromagnetics are, of course, not possible. In many materials the forces opposing rotation are much greater than those opposing wall movement, and in single-domain particles of such materials we shall expect much lower permeabilities and larger coercive forces than in the normal material. As an example we may consider a small sphere of iron to which a magnetic field is applied in a (100) plane in a direction making an angle θ with the

[100] axis (fig. 53 a). The magnetocrystalline energy will depend on the direction of magnetization of the sphere, according to equation (2.1). Its variation with angle in the (100) plane is shown in fig. 53 b. In zero field the magnetocrystalline energy is the only factor affecting the direction of magnetization, which will accordingly lie in one or other of the minima of the upper curve of fig. 53 b. The application of a field H in the direction θ will add an energy term—the magnetic potential energy of the sphere in the field—which will alter the energy curve as shown by the lower lines in fig. 53 b. It is clear that if the magnetization is initially in the energy minimum [010], it will be turned towards the field direction, reversibly in small fields, but by an irreversible jump to the minimum [100] if the field exceeds a certain critical value. This critical field, whose order of magnitude can be estimated as K_1/I_s, is the coercive force for the single sphere, and its average over all the particles in a ferromagnetic material made up of single-domain particles should give the coercive force of the material.

The magnetocrystalline term is only one of the possible sources of anisotropy in the single-domain particles. The other ones likely to be important are anisotropy caused by stresses, and anisotropy of shape. This last effect is the one considered in most detail by Stoner and Wohlfarth and shown by them to be the most probable source of very high coercivities (> 400 oersteds). If the single-domain particles are not spherical, then their demagnetizing fields, and consequently their energy, will depend on the direction of magnetization, being least when they are magnetized along the longest axis (direction of least demagnetizing coefficient). By plotting energy curves, corresponding to those of fig. 53 b, Stoner and Wohlfarth deduced the critical fields for particles of various dimension ratios and orientations with respect to the field, assuming that they had the form of ellipsoids. For a collection of prolate spheroids oriented at random they obtained $H_c = 0\cdot479(N_b - N_a)\,I_s$, where N_a and N_b are the demagnetizing coefficients along the polar and equatorial axes. If the dimension ratio is $1\cdot1$, $N_b - N_a$ is $0\cdot472$, and for iron, with $I_s = 1720$ gauss, H_c becomes 390 oersteds. These results agree closely with those of Néel for a similar model.

6.6. Other types of inclusion: stress centres

In the last two sections we have considered heterogeneous materials containing two sharply distinct phases, the ferromagnetic and the non-ferromagnetic. The essential feature is, however, the fluctuation of I_s, and this may occur within a ferromagnetic—for instance, through fluctuations of composition of an alloy—without the actual occurrence of a non-ferromagnetic phase. A second cause of variation of I_s—in direction, not in magnitude— is the presence of internal stresses. In our earlier discussions of the effects of stresses, we have taken into account only the component of stress parallel to the magnetization. If, however, there is a region within any domain where the stress tends to rotate the magnetization from its average direction, then divergence of I_s will occur at the boundaries of the region and a distribution of free poles similar to that around an 'inclusion' will appear, as shown in fig. 54.

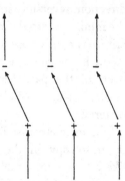

Fig. 54. Free poles near a region where magnetization deviates from its mean direction.

It has been pointed out by Néel that there is considerable energy in the demagnetizing field associated with such a distribution of free poles round a stress centre and that this energy is reduced by redistribution of the free poles if the stress centre is bisected by a domain wall, just as in the case of an actual inclusion. A definite energy must therefore be supplied by a magnetic field in order to move a wall from a position bisecting a stress centre, and this effect of stresses in hindering wall movement must be taken into account in addition to their hindering effect through variations in wall energy considered in § 6.3. A rough calculation of energies, from a model like that of fig. 54, suggests that the two types of hindrance to wall movement are of the same order of magnitude for a given internal stress variation, but Néel, developing the stresses throughout a material as a Fourier series, concludes that the effect resulting from divergence of magnetization may be much more important than that due to variation of wall energy. It does not seem possible at present to say with certainty which mechanism predominates in any particular case.

6.7. Applications to real materials

The results obtained in the last few sections do not by themselves make it possible to calculate the magnetic properties of materials in terms of their mechanical state. They were, for the most part, calculated for very simple models of isolated domains and isolated irregularities of stress or composition. In order to apply them to real materials, it is necessary to assume some plausible distribution of the irregularities throughout the material and then inquire how the various effects already considered in isolated cases will affect the domain arrangements in the whole material. It is clear that the detailed magnetic properties—the exact shape of the magnetization curve, for instance—will depend on the precise distribution of the irregularities, so that agreement in detail between theory and experiment is not necessarily to be expected so long as mathematically simple forms of the irregularities are assumed. If sufficiently complicated assumptions are made, it is possible to interpret almost any magnetization curve, but this is not a very profitable line of investigation; a theory can be accounted satisfactory if it explains the values of certain easily specified constants, such as the initial permeability and the coercive force, in terms of a reasonably small number of parameters.

6.8. The Becker-Kersten stress theory

The first attempt to explain observed magnetic properties in terms of inhomogeneities in the material was that of Becker and Kersten, who took account of the effects considered in §§ 6.2 and 6.3, the influence of internal stresses on the volume energy of domains and on the surface energy of their walls.

The simple calculations of §§ 6.2 and 6.3 deal with single walls, but they can at once be extended to cover the behaviour of a material with many walls if the stress and gradient of stress are known at every point on every wall, or if suitable mean values are given. Kersten made the simple assumption that at every point the magnetization had the direction made most favourable by the local stress, so that in a material with alternate compression and tension there would be a 90° wall at every zero of stress (fig. 47). This assumption can be criticized in the light of the ideas about domain arrangements presented in Chapter IV; it is obvious, for example, that it would be energetically unprofitable to have a wall at every zero of a small stress alternating with short wave-length. Nevertheless, the assumption leads to simple results which are a useful basis for discussion. Equation (6.1) gave the change of magnetization due to the movement of unit area of a single wall in a field H as

$$\Delta I = \frac{I_s^2}{a\lambda\sigma_i} H,$$

where a was the wave-length of a sinusoidally varying stress of amplitude σ_i. If there is a wall for every zero of σ_i, then the total number of walls in unit length is $1/a$ and the change in the magnetization of unit volume is

$$\Delta I = \frac{I_s^2}{\lambda\sigma_i} H.$$

We can therefore write the initial susceptibility due to the movement of 90° walls as

$$\chi_{90°} = \frac{\Delta I}{H} = \frac{I_s^2}{\lambda\sigma_i}.$$

More detailed calculation, allowing for randomness of the relative orientation of stresses and walls, introduces a factor $4/3\pi$, that the initial susceptibility becomes

$$\chi_{90°} = \frac{4}{3\pi} \frac{I_s^2}{\lambda\sigma_i}. \qquad (6.7)$$

If 180° walls were present in addition to the 90° ones, their movement could contribute to the initial susceptibility, the magnitude of their contribution being calculable as follows. A 180° wall will be in equilibrium at a point where the rate of change of wall energy with position, $\partial\gamma/\partial x$, is equal to the rate of change of magnetic energy, $2HI_s$. A change in field δH will thus cause a wall movement $\delta x = \frac{2\delta H I_s}{\partial^2\gamma/\partial x^2}$ and a change in magnetization

$$\Delta I = 2I_s\delta x = \frac{4I_s^2}{\partial^2\gamma/\partial x^2} \delta H.$$

If we assume a sinusoidal variation of γ with position, we can write

$$\frac{\partial^2 \gamma}{\partial x^2} = \frac{2\pi}{l} \frac{\partial \gamma}{\partial x},$$

where l is the wave-length of the stress variation. An estimate of $\partial \gamma / \partial x$ can be deduced from equation (6.2) or (6.3) as was done in § 6.3, giving $\partial \gamma / \partial x \approx \lambda \sigma_i$. We thus obtain the expression

$$\chi_{180°} \approx \frac{2 I_s^2}{\pi \lambda \sigma_i} \qquad (6.8)$$

for the susceptibility caused by 180° wall movement, assuming that there is a wall at every zero of γ and therefore $1/l$ walls in unit length.

This assumption, however, cannot be justified by reference to magnetostrictive energy, as could the analogous assumption for 90° walls. Kersten was, in fact, able to deduce from experimental results that in many cases only a small fraction of the possible sites for 180° walls were actually occupied by walls, so that their contribution to the initial susceptibility was much less than suggested by equation (6.8), and could be neglected in comparison with the contribution from 90° wall movement. The experimental evidence for the absence of a 180° wall contribution is the observation of χ_r, the reversible susceptibility at the remanence point, and its ratio to the initial susceptibility, χ_0. The mechanism of magnetization determining χ_r is very similar to that determining χ_0, namely, the reversible movement of domain walls in very small fields, but whereas both 90° and 180° walls (if present) contribute to χ_0, χ_r can be assumed to be due to 90° walls only, all 180° walls being removed by the process of reaching the remanence point. The ratio χ_r / χ_0 can therefore be used to assess the contribution of 180° wall movement to χ_0. Even if 180° walls do not contribute at all to χ_0, the expected value of the ratio χ_r / χ_0 is not unity, because the orientation factor in equation (6.7) is different in the two cases. Becker obtains the value $1 - 2/\pi$ for χ_r / χ_0 on the assumption that both χ_r and χ_0 result from 90° wall movement alone. If 180° walls contribute to χ_0, the ratio will be smaller. In many materials χ_r / χ_0 is found to be about 0·3, indicating that the contribution of 180° walls to χ_0 is small.

The calculations just made for the initial susceptibility refer to materials in which the magnetocrystalline anisotropy is much greater than that resulting from internal stresses, so that the directions of domain magnetization are very nearly the easy directions of the magnetocrystalline effect. The other extreme case, where the internal stresses are so large, compared with the magnetocrystalline anisotropy, as to be the chief factor determining the direction of magnetization, was also considered by Becker and Kersten. They assumed that the internal stress had a constant

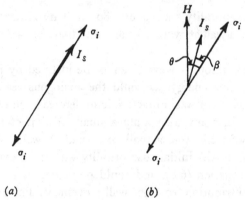

(a)   (b)

Fig. 55. Rotation of domain magnetization, I_s, away from direction
of stress, σ_i, by a field, H.

magnitude, σ_i, but varied in direction in a random manner. In zero field the direction of magnetization was supposed to be the same as the direction of σ_i at every point. The effect of a small field making an angle θ with the direction of the stress is to turn the magnetization towards itself, as shown in fig. 55, and thus increase the magnetization in the field direction. The angle, β, through which the magnetization turns can be found by equating the couple due to magnetostrictive forces to the couple due to the field H. The former is, by equation (3.20), equal to $3\lambda\sigma_i \sin \beta \cos \beta$, and the latter is $HI_s \sin \theta$, if β is very small. We thus obtain

$$\beta = \frac{HI_s \sin \theta}{3\lambda\sigma_i}.$$

The corresponding change in the component of I_s in the direction of H is

$$\Delta I = I_s \sin \theta . \beta,$$

and we obtain for the initial susceptibility in small fields

$$\chi_0 = \frac{I_s^2}{3\lambda\sigma_i} \sin^2\theta.$$

The average of $\sin^2\theta$ over a sphere is $\frac{2}{3}$, so that the final result for the initial susceptibility of a material subject to randomly oriented internal stresses is

$$\chi_0 = \frac{2}{9} \frac{I_s^2}{\lambda\sigma_i}. \tag{6.9}$$

It is remarkable that this result for materials with large internal stresses should be so nearly equal to that obtained in equation (6.7) for materials with small internal stresses.

Other quantities that can be evaluated from the Becker-Kersten theory are the effect of small external stresses on remanence, and, in the case of materials with large internal stresses, the 'work of magnetization' from the unmagnetized state to saturation.

It is clear that the order of magnitude of the first effect will be given by

$$\frac{\partial I_r}{\partial\sigma} \approx I_s \frac{\sigma}{\sigma_i},$$

where σ and σ_i are the (uniform) external stress and (random) internal stress respectively. Allowance for the effects of different orientations introduces a numerical factor $\frac{1}{4}$, so that the final result is

$$\left(\frac{\partial I_r}{\partial\sigma}\right)_{\sigma\to0} \approx \tfrac{1}{4} I_s \frac{\sigma}{\sigma_i}, \tag{6.10}$$

an expression which applies approximately to both extreme cases, the one where domain magnetizations are controlled chiefly by magnetocrystalline anisotropy as well as the one where they are controlled by the magnetostrictive effects of large internal stresses.

The 'work of magnetization' is measured by the area between the I axis and the magnetization curve. (If hysteresis effects are present, the reversible work of magnetization is obtained by using only the descending branch of the hysteresis loop.) It represents the work done by the magnetic field in turning the domain magnetizations, against whatever forces constrain them to their positions in the unmagnetized material, to their saturated position of complete alignment with the field. In materials where stress effects far outweigh magnetocrystalline ones the magnetization will be random with respect to the crystal axes both in the initial and in the final state. The change in magnetocrystalline energy is therefore zero, and the work of magnetization is entirely expended in turning magnetizations from positions where their stress energy is zero to their final position in which the internal stresses are oriented quite at random to the direction of magnetization. The work of magnetization, U, is therefore given by the mean over all θ's of the energy of a domain magnetized at angle θ to the direction of a stress σ_i. From equation (3.20) we can write this stress energy as $\frac{3}{2}\lambda\sigma_i\sin^2\theta$, and the mean value over all orientations is simply

$$U=\lambda\sigma_i \tag{6.11}$$

Although, according to the Becker-Kersten theory, the 180° walls play only an insignificant part in changes occuring in low fields, they are all-important in deciding the coercive force, since reversal of magnetization cannot take place without the movement of such walls. The effects of stress variations on 180° walls were considered in § 6.3, and expressions were obtained for the variation of wall energy with position. The field required to force a wall past a stress obstacle for which $\partial\gamma/\partial x$ has the maximum value $(\partial\gamma/\partial x)_{\text{max.}}$ is

$$H_0=\frac{1}{2I_s\cos\theta}\left(\frac{\partial\gamma}{\partial x}\right)_{\text{max.}}, \tag{6.12}$$

where θ is the angle between the directions of the magnetization, $\pm I_s$, and the field, H_0. The coercive force for a material will be some suitable mean of H_0 for all its walls; Becker, using equation (6.3) for $(\partial\gamma/\partial x)_{\text{max.}}$ and allowing for variation of θ, obtains

$$H_c=\tfrac{3}{2}\lambda\sigma_i\,\frac{1}{I_s}\frac{\delta}{l}\quad(\delta\ll l),$$

$$H_c=\tfrac{3}{2}\lambda\sigma_i\,\frac{1}{I_s}\frac{l}{\delta}\quad(\delta\gg l).$$

The maximum value of the coercive force is thus

$$H_c\sim\tfrac{3}{2}\lambda\sigma_i\,\frac{1}{I_s},\quad\text{when}\quad\delta\approx l. \tag{6.13}$$

The identification of H_c with a mean value of H_0 presupposes that the reduction of magnetization from remanence to zero is, in fact, brought about by the movement of 180° walls. In an idealized material there would be no 180° walls present at remanence, and we should have a situation like that considered in § 6.5, where reversal of magnetization could only occur through rotation of whole domains against anisotropy or stress forces. Rotation against stress forces actually leads to coercive forces of the same order as in equation (6.13), but rotation against magnetocrystalline anisotropy requires much larger fields, and we must assume that in most materials there are irregularities which act as centres from which 180° walls can spread and cause reversal of magnetization for the lower coercive forces given by equations (6.6) and (6.13).

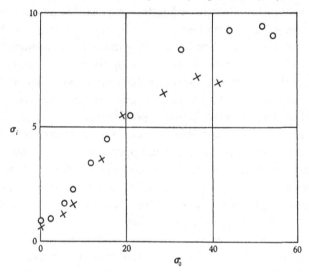

Fig. 56. Internal stress, σ_i, in a nickel wire after plastic stretching by a stress σ_0 (σ_i and σ_0 in Kg./mm.²). Circles show points deduced from initial permeability measurements; crosses show points deduced from measurements of change in remanence by stress. (After Kersten, 1933.)

6.9. Experiments on the stress theory

It would be desirable to test the conclusions of the Becker-Kersten theory by comparison of magnetic measurements with results predicted from equations (6.7)–(6.13), using values of σ_i obtained by some independent method. Unfortunately, no satisfactory method of measuring σ_i has been suggested so far, so that almost the only checks of the theory are 'internal' ones, of the consistency of values deduced by different magnetic methods.

One 'external' check of some value is provided by the way in which magnetic properties depend on plastic deformation, which may be supposed to increase the internal stresses. Kersten (1933) subjected nickel wires to various large stresses σ_0, sufficient to stretch them plastically, and afterwards measured the initial susceptibility, χ_0, and the effect of small applied stresses on the remanence, $\partial I_r/\partial \sigma$. From each of these quantities, σ_i could be found using equations (6.9) and (6.10) respectively. The results are shown in fig. 56 as graphs of the values obtained by the two methods against the previously applied stress, σ_0. It will be seen that the two methods lead to very similar values of σ_i, and that these bear an almost linear relation to σ_0, $\sigma_i = \frac{1}{4}\sigma_0$, for stresses below 20 Kg./mm.².

These experiments studied the variation of χ_0 when σ_i was altered. An alternative line of investigation is the dependence of χ_0 on λ, the other variable of equation (6.9). It has been known for a long time that alloys with low magnetostriction have high permeabilities, as equation (6.9) would suggest, though the exact properties of each alloy are, of course, dependent on its heat treatment. The effect of the heat treatment is to vary the internal stresses and hence χ_0. A lower limit to σ_i and an upper limit to χ_0 is, however, set by stresses of magnetostrictive origin which must be set up as the material is cooled through its Curie point. Kersten (1931 b) has estimated the magnitude of these stresses and hence the maximum value of χ_0 in a given alloy, and has compared the predicted maximum values with those reached in practice. He assumes simply that the increase in domain magnetization on cooling results in a change in dimensions of order λ, and that stresses of the order of λE, where E is Young's modulus for the material, are set up. On substituting $\sigma_i = \lambda E$ in equation (6.9) we obtain

$$\chi_0 = \frac{2}{9}\frac{I_s^2}{\lambda^2 E}. \tag{6.14}$$

Values calculated from this equation are shown in fig. 57, together with the highest values measured experimentally in 1931. It has since proved possible to exceed these values of χ_0 by special treatments (such as annealing in a magnetic field) which produce non-random orientation of the internal stresses, and so falsify the assumptions on which equation (6.9) is based.

Other 'internal' tests of the theory, comparing values of σ_i deduced from different magnetic measurements, may be mentioned briefly. The expression (6.11) for the work of magnetization has been used by Kersten (1931 a) to find σ_i in a nickel wire of which he had also measured the initial susceptibility. The reversible work of magnetization was 25,000 ergs/c.c. and the initial susceptibility was 2·5. Equations (6.11) and (6.9) give $\sigma_i = 7$ Kg./mm.^{2} and

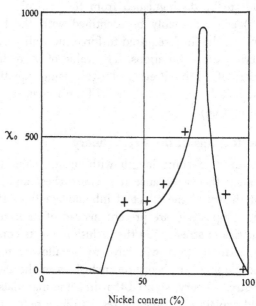

Nickel content (%)

Fig. 57. Maximum values of initial susceptibility, χ_0, for iron-nickel alloys of various compositions. Curve shows observed values, crosses show values calculated from magnetostriction measurements, by equation (6.14). (After Kersten, 1931 b.)

$\sigma_i = 6$ Kg./mm.^{2} respectively, which is remarkably close agreement. Measurements and comparisons of χ_0, $\partial I_r/\partial\sigma$ and U have also been made, among others by Förster and Stambke (1941), and their results agree well with the theory.

The effect of internal stresses on the movement of 180° walls is covered by equations (6.2) and (6.3). These equations give values of H_c agreeing as to order of magnitude with the coercive forces which actually occur, but more precise comparison with theory can be made if the 'coercive force' of a single 180° wall is measured, rather than the coercive force of a large piece of material containing many walls. This can actually be done by experiments like those of Sixtus and Tonks's (see § 5.7), in which reversal of magnetization in a wire under tension occurs by the movement of a 180° wall along the wire. The law governing the speed of movement, V, is found to be $V = A(H - H_0)$, where A is a constant, H the applied magnetic field and H_0 a critical field below which propagation of the wall will not take place. This critical field (to be distinguished from the critical *starting* field H_s of § 5.7) can reasonably be identified with the field H_0 of equation (6.12), the field required to force the wall past obstacles due to internal stress variations. The value of σ_i deduced from measurements of H_0 by Preisach (1932), using equation (6.12), agrees quite closely with that deduced from measurements of χ_0, using equation (6.9).

6.10. Imperfections of the stress theory

We have dealt at some length with the development of the Becker-Kersten theory because it is the earliest and most fully worked-out theory of the effect of inhomogeneities. If inhomogeneities of composition are present instead of or in addition to inhomogeneities of stress, then the inclusion effects considered in elementary form in § 6.4 will have to be allowed for as well. Kersten, developing the theory on inclusions in the same sort of way as the stress theory, showed that in some materials inclusion effects could explain experimental results much better than could the effects of stress.

There are, however, grave objections to all such calculations of the effects in extended materials of the 'elementary' hindrances to wall movement considered in §§ 6.2–6.6. The fundamental objections are that no account is taken of the requirements outlined in Chapter IV, which all domain structures must satisfy, and that the distributions of obstacles which are used, implicitly or explicitly, in the development of the theories are so far from being irregular that they give entirely false results.

These points can be illustrated most simply by considering the 'inclusion' type of obstacle, though they apply also to 'stress obstacles'. Kersten's model of a wall in a material containing inclusions was shown in fig. 50. The calculations are all made on the assumption that the inclusions are distributed regularly, at the corners of a cubic lattice, so that a plane wall intersects many inclusions at the same distance from their centres and the hindering effects of all these inclusions are additive. It is clear that these calculations will not apply at all to an arrangement with random distribution of inclusions. In such a case the inclusions intersected by a large plane wall would not all lie on the same side of the wall, and the effect of those with centres to the left of the wall, tending to move it to the left, would roughly cancel the effect of those with centres to the right, tending to move it to the right. The residual effect, resulting from statistical fluctuations in the number of inclusions to left and right of the wall as it moves, can be estimated very approximately as of the order of the effect calculated for a regular array of inclusions divided by the total number of inclusions intersected by the individual wall. More detailed calculations by Néel (1946) confirm this estimate, showing that the maximum coercive force in a typical case is reduced from the value of several hundred oersteds calculated by Kersten for a regular array to one of a few oersteds, by allowing for the irregularity of the distribution of inclusions.

This first point, that inclusions are unlikely to be regularly spaced in the way assumed by the theory, is only important in conjunction with the second point, that domains and their walls have to be arranged according to certain rules discussed in Chapter IV. If a domain wall were free to take up any shape—the one with minimum area in particular—it would be deformed in such a way as to intersect as many inclusions as possible, and the forces opposing its movement would be approximately the same whether the inclusions were regularly or irregularly distributed. It is only because local demagnetizing fields, set up by the free poles appearing at a deformed domain wall, tend to keep the wall very nearly plane, that the argument for the averaging out of the effects of different inclusions is valid.

If we allow for the fact that a wall is not a perfectly rigid plane but one that can be deformed slightly against the action of internal demagnetizing fields, it is clear that the averaging out of the effects of different inclusions will not be complete. The actual behaviour of the wall will thus be intermediate between that with a regular array of inclusions and that with a rigidly plane wall and randomly arranged inclusions.

In the development of the stress theory (§ 6.8) the hypothesis of regular distribution of the obstacles to wall movement is not stated quite so explicitly as in the inclusion theory, but is implied by the one-dimensional treatment, which assumes that the 'obstacles' are uniform in the plane of the wall. Néel has shown that with an irregular distribution of stresses the maximum coercive force that can be caused by the effects of stress on wall energy is only of the order of 1 oersted if the wall is treated as a rigid plane or 10 oersteds if allowance is made for deformation of the wall (for materials with wall energies of the order of 1 erg/cm.).

6.11. Néel's theory

Néel concludes, therefore, that the mechanisms outlined in §§ 6.3 and 6.4 cannot provide large enough obstacles to wall movement to account for the coercive forces of several hundred oersteds which are observed in many materials. He suggests that the chief obstacles to wall movement are those caused by the mechanism of § 6.6, involving the strong tendency of walls to bisect 'islands' of free poles created by fluctuations of the domain magnetization. The analysis of the effects of a random arrangement of such obstacles is complicated, but Néel's final result for the coercive force may be quoted.

The coercive force which is due to islands of free poles caused by internal stresses with stress energy C ($= \frac{3}{2}\lambda\sigma_i$ if the magnetostriction, λ, is isotropic), which affect a fraction v of the total volume of the material, is

$$H_c \approx \tfrac{4}{15}\pi \frac{vC^2}{KI_s}\left[1\cdot386 + \log\sqrt{\left(\frac{2\pi I_s^2}{K}\right)}\right] \approx \frac{1}{4}\frac{vC^2}{KI_s}$$

for materials in which C is small compared with the magnetocrystalline energy K, and

$$H_c \approx 0\cdot69\,\frac{vC}{I_s}\left[1\cdot386 + \log\sqrt{\left(\frac{6\cdot8I_s^2}{C}\right)}\right] \approx \frac{vC}{I_s}$$

for materials with $C \gg K$.

The coercive force which is due to islands of free poles caused by non-magnetic inclusions occupying a fraction v' of the total volume is

$$H'_c = \frac{2Kv'}{\pi I_m}\left[0\cdot 386 + \log\sqrt{\left(\frac{2\pi I_m^2}{K}\right)}\right] \approx \frac{Kv'}{I_m},$$

where I_m is the mean of I_s over ferromagnetic and non-ferromagnetic. If the appropriate constants for iron and nickel are substituted in these formulae, it is found that coercive forces of the order of several hundred oersteds can be attained in each metal. The approximate equations, assuming reasonable maximum values for σ_i of 3×10^9 dynes/cm.², are

$$H_c \approx 2\cdot 2v + 360v' \quad \text{for iron} \quad (C \ll K),$$

and $\qquad H_c \approx 330v + 97v' \quad \text{for nickel} \quad (C \gg K).$

These results show that high coercive forces in iron-like materials must be attributed to the effect of inclusions, but that in nickel-like materials they must be due to internal stresses.

It is clear from Néel's treatment of the effects of irregular distributions of stresses and inclusions that the degree of hindrance to wall movement depends very markedly on the 'shape' and arrangement of the variations in stress and composition. An arrangement with sharp stress maxima would, for instance, provide much more hindrance than one with the same stress varying more smoothly. Moreover, any trace of regularity (for example, a tendency for stresses to congregate on parallel crystal planes) could enormously increase the calculated effects. The 'random' arrangement considered by Néel is not the only conceivable one; it has certain characteristics of peakiness and regularity which are fixed by the coefficients he assumes in the Fourier development of the inhomogeneities. It is by no means certain that these characteristics agree with those of the 'random inhomogeneities' of actual materials, so that Néel's estimate of the relative importance of the various types of obstacle and the actual values of coercive force resulting from each are not necessarily correct.

6.12. Modification of the stress theory

Néel's calculations have not yet been extended to cover the effects of random irregularities in the low-field region, where the

Becker-Kersten theory was able to correlate so many facts. It seems possible, however, to retain many of the formal results of the older theory if σ_i is reinterpreted as an 'effective' internal stress, defined, for example, by equation (6.7) and measuring not the actual internal stress but simply the amount of hindrance to wall movement provided by stress obstacles. The relation between this effective σ_i and the actual value of the internal stress will depend on such things as the area and rigidity of the domain walls, the number and sharpness of the stress obstacles affecting individual walls, and the degree to which the movements of different walls are linked by structural requirements. It is even conceivable that in an entirely stress-free material the effective σ_i might have a finite value, representing the resistance to any wall movement that involves increases in the area of domain walls or the volume of closing domains. This might be called the 'structural' component of σ_i. Its importance will be small unless the 'structural energy', W_s (as calculated for a particular example in § 4.6), is comparable with the stress energy (more strictly, unless the change in W_s for a given change in magnetization is comparable with the change in stress energy). In most cases W_s is not greater than a few hundred ergs/c.c. and can be neglected in comparison with the normal $\lambda\sigma_i$ values of at least 1000 ergs/c.c.*

Although the effective σ_i which must now be used in the application of the stress theory may be very different from the actual stress value, most of the applications of the Becker-Kersten theory were comparisons of materials that would be expected to have roughly similar domain arrangements, so that the order-of-magnitude agreement of σ_i values deduced from different experimental methods will still be preserved. If, however, there are two specimens with similar internal stresses but, for some reason, different domain arrangements, then the 'effective' internal stresses will not be the same and the simple stress theory can give misleading results. An example is provided by the dependence of initial susceptibility on crystal orientation.

* It has recently been shown by Lee (1953) that W_s has quite important effects on the magnetization curves of single crystals in low fields; the chief effect is a rounding of the abrupt knee suggested by the simple calculations of Chapter II. Good agreement is found with the experiments of Williams (fig. 59).

6.13. Directional dependence of initial susceptibility

According to the stress theory the initial susceptibility of a crystal can be calculated in terms of the internal stress variations, from a model such as that in fig. 47. In discussing this model we considered only one of the three possible types of wall (those in the y-z plane) and only one of the infinite number of possible axes of stress variation (the x-axis). When these various possible orientations are taken into account, as was done by Becker, the factor $4/3\pi$ is introduced into equation (6.7), and it is found that the initial susceptibility is independent of the orientation of the magnetic field with respect to the crystal axes, for cubic crystals. This result, that χ_0 is isotropic, was also demonstrated by Becker in a simpler and more general way. The only essential assumption is that the three principal axes of the crystal are exactly equivalent magnetically. If χ_{100}, χ_{010} and χ_{001} denote the initial susceptibility measured in these three directions, we can then write

$$\chi_{100} = \chi_{010} = \chi_{001} = \chi.$$

The components of magnetization along the three axes when a small field, H, is applied in a direction given by direction cosines α_1, α_2, α_3 will be

$$I_{100} = \chi_{100} H \alpha_1, \quad I_{010} = \chi_{010} H \alpha_2, \quad I_{001} = \chi_{001} H \alpha_3,$$

and the resultant magnetization is

$$I_{(\alpha_1 \alpha_2 \alpha_3)} = H[\chi_{100}\,\alpha_1^2 + \chi_{010}\,\alpha_2^2 + \chi_{001}\,\alpha_3^2]$$
$$= H\chi$$

in the direction of H, provided that the field is so small that the I-H relation is linear. The initial suceptibility is thus independent of orientation in the crystal.

This result does not, however, agree with experiment. The measurement of χ in single-crystal specimens is difficult because the specimens are usually small and have large demagnetizing coefficients. Accurate measurements have, however, been made by Williams (1937b), using single crystals cut as shown in fig. 58 so that a closed magnetic circuit is provided, with the flux almost everywhere parallel to one particular crystal axis. The results shown in fig. 59 indicate considerate anisotropy in χ. It follows that the assumption of equivalence of the three principal crystal axes must be false. Kondorsky (1938) suggested that the experiments could be explained by assuming that in small fields the domains had a preference for those easy directions which lay parallel to the surfaces of the specimen; this 'preferential orientation' of domains would alter the orientation factor in equation (6.7),

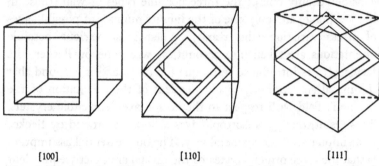

[100]   [110]   [111]

Fig. 58. Method of cutting specimens from cubic single crystals to give closed magnetic circuits in the principal directions. (After Bozorth, 1937.)

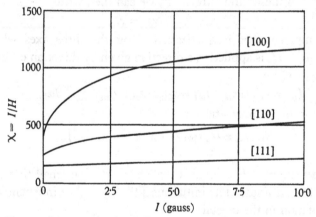

Fig. 59. Susceptibility, χ, of Williams's (1937b) specimens as a function of intensity of magnetization, I.

giving the ratio $\chi_{100} : \chi_{110} : \chi_{111} = 6 : 3 : 2$

for the specimens used by Williams, a ratio close to that found in the experiments.

Further evidence for the preferred orientation of domains in low fields is provided by experiments of Shoenberg and Wilson (1946). They measured the torque on flat single-crystal disks in magnetic fields too low to cause appreciable turning of domains away from the easy directions as a function of the angle between field and crystal axes. If the three principal axes were equivalent, the torque on a disk whose plane is (110) should be zero; in fact, an appreciable component of torque with period 180° was found, indicating a preference of domains for the [100] and [$\bar{1}$00] directions, lying in the plane of the disk, rather than the [0 ± 10] and [00 + 1] directions which project out of the plane at angles of 45°.

The reason for this preferred orientation of domains can readily be understood if we consider the actual domain structure of the crystal. If the main structure consists of domains with magnetization in the easy directions parallel to the surfaces of the specimen, then no 'closing domains' will be needed (except near the edges of the disk). If, however, the main structure had its domains magnetized in the easy directions projecting out of the plane of the specimen, then closing domains would be needed to avoid the appearance of free poles on the surface, as discussed in § 4.3. The energy associated with these closing domains is avoided in the arrangement with domains parallel to the surfaces, which will thus be the one normally occurring.

6.14. Formal treatments of the Rayleigh region

We have shown that stresses and inclusions can produce obstacles to domain wall movement, but that it is not possible to build up a satisfactory detailed picture of the behaviour of a ferromagnetic in terms of these obstacles. An alternative approach is, however, possible, in which a more formal theory is sought, ignoring the question of the origin of obstacles, but considering the macroscopic magnetic effects that would be expected if obstacles existed and had sizes and positions fixed by some simple law. The aim of such a treatment is to explain the form of a magnetization curve or of the relation between magnetic quantities, rather than the absolute value of a coefficient such as the initial susceptibility; the aim of the treatments outlined in earlier sections was to obtain actual values of easily definable coefficients, while leaving undecided such questions as the form of the magnetization curve in finite fields. A complete account of magnetic behaviour would, of course, have to combine both treatments, but this has not yet been done.

The familiar hysteresis loops of technical materials have such various and complicated shapes that there is little hope of interpreting them by a formal theory, without introducing a large number of parameters. It was, however, found by Rayleigh (1887) that in very low fields the hysteresis loops of many materials approximate to a simple common form shown in fig. 60, the equation being

$$B = (\mu_0 + \alpha H_1)\, H \pm \tfrac{1}{2}\alpha(H_1^2 - H^2), \qquad (6.15)$$

where μ_0 and α are constants of the material, H_1 is the maximum

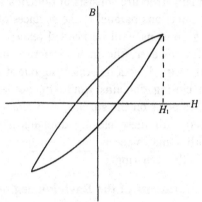

Fig. 60. Form of hysteresis loop found in low fields.
(After Rayleigh, 1887.)

field used in traversing the loop, and the positive and negative signs correspond to the descending and ascending branches of the loop respectively. Rayleigh's results were obtained on various specimens of soft iron in the range of H_1 between 2×10^{-5} and 4×10^{-2} oersted; it has since been shown (e.g. Ellwood, 1935) that equation (6.15) holds for other ferromagnetics in sufficiently low fields.

In this very-low-field region of magnetization the relations are thus so simple as to offer some hope of explanation by a formal theory that is not unreasonably complicated. Such theories have been put forward by Preisach, Kondorsky, and Néel. Preisach (1935) studied the properties of an assembly of domains each with a rectangular hysteresis loop as shown in fig. 61. Each domain could be characterized by the constants a and b of its loop, the displacement field, b, representing the action of neighbouring domains. He found that if all values of a and b were assumed to be equally probable, a parabolic relation between field strength and magnetization, as required by Rayleigh's laws, was obtained. The reversible part of the magnetization, specified by μ_0 in equation (6.15), was not covered by this mechanism. The method was extended, more account being taken of the interactions between domains, by Kondorsky (1942).

This interpretation of the Rayleigh loop is satisfactory in a purely formal way, but the rectangular loops assumed for the domains do not correspond at all closely with the way we should expect

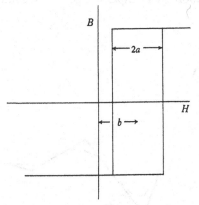

Fig. 61. Preisach's hypothetical elementary hysteresis loop.

domains to behave with any plausible type of 'obstacle' resulting from internal strains or inclusions. A new approach was made by Néel (1942). He considered the changes of magnetization brought about by the movement of a domain wall among a series of obstacles, the effect of the obstacles being represented by a 'characteristic function' giving the energy of the system as a function of the position of the wall. As a first approximation, he assumed the form of the characteristic function to be a series of straight lines (fig. 62), the slopes of the lines being distributed at random according to a Gaussian probability law. As a second approximation, the straight lines were replaced by parabolic arcs, giving a physically more plausible shape to the characteristic function. If the movements of all the walls in a material are controlled by such functions, Néel showed that the reversible and irreversible changes of magnetization would be those described by Rayleigh's relation.

6.15. Practical applications

In this chapter we have shown that variation in internal stress or in composition in a ferromagnetic can produce 'obstacles' to the movement of its domain walls in several ways. It seems likely that such obstacles can account for many of the properties of ferro-magnetics that are not covered by the laws applying to ideally homogeneous materials, but the phenomena are so complicated that the detailed prediction of properties is not yet possible. Theory, in fact, does little more than emphasize the long-familiar practical idea

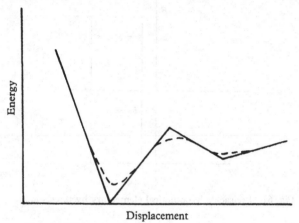

Fig. 62. Néel's (1942) representation of energy changes accompanying wall movement.

that materials with small obstacles, i.e. with low internal stresses and small amounts of impurity, will tend to have high permeability and low coercive force, because their walls can move freely, while materials with large stresses or impurity contents will have low permeabilities and high coercive forces.

The practical applications of modern ideas on domain behaviour to technical magnetic material have recently been summarized by Brailsford (1948a) and by Hozelitz (1952). A good example is the development of the high-permeability ferrites by Snoek (1947) and his co-workers. Snoek pointed out that the highest permeability would be expected in materials where both the magnetostriction coefficient, λ, and the magnetocrystalline one, K, were small, thus reducing both the obstacles due to internal stresses (§ 6.3) and those due to inclusions (§ 6.4). These coefficients can be varied by varying the composition of an alloy; there are several binary alloys for which one or other of the coefficients becomes zero at a particular composition, but in general it is only by using an alloy with three components that the two can be made to vanish simultaneously. Fig. 63 shows the variation of K in the system Fe-Al-Si and the lines of zero K and λ. The alloy corresponding to

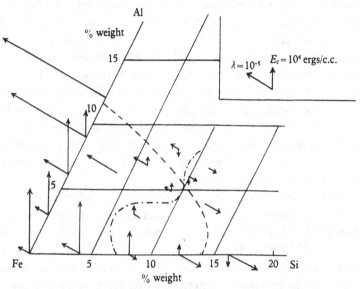

Fig. 63. Variation of magnetostriction and magnetocrystalline anisotropy with composition in Fe-Al-Si alloys. Lengths of arrows represent values of magnetostriction, λ (in direction of minimum energy), and anisotropy energy, E_c (corrected for effect of internal stresses). Dotted lines are zero lines of λ ($-\cdot-\cdot-$) and E_c ($----$). (After Snoek, 1947.)

the intersection of these lines is 84 % Fe, 6 % Al, 10 % Si, which is the high-permeability alloy 'Sendust', discovered in Japan. In the case of the ferrites ($Fe_2O_3 . M^{++}O$, where M is any bivalent metal), Snoek found that the magnetostriction and the magneto-crystalline anisotropy could be controlled by changing the proportions of ions in 'mixed ferrites' ($Fe_2O_3 . nM_1^{++}O . (1-n)M_2^{++}O$), and that both could be made to vanish simultaneously, giving the new high-permeability material 'Ferroxcube'.

The oldest high-permeability material is the iron-nickel alloy with about 78 % nickel, known as 'Permalloy' or (with small additions of copper) 'Mumetal'. The alloy with 82 % nickel has zero magnetostriction coefficient, and the magnetocrystalline coefficient also passes through zero at about this composition. It has long been a puzzle that the maximum of permeability does not exactly coincide with the minimum of magnetostriction on the composition curve. It seems likely that for the highest permeability a compromise between minimum magnetostriction and minimum magnetocrystalline anisotropy is necessary. (It has been shown by Grabbe (1940) that suitable heat treatment can reduce the constant K to zero in nickel-iron alloys of the Permalloy type.) The details of the 'Permalloy problem' are, however, not yet completely cleared up.* 'Domain theory' for these materials is, in any case, very vague, because the width of domain walls, when the values of λ and K are so small, becomes enormous. With plausible values of λ and K the wall width in high-permeability materials, as indicated by equations (5.15) and (5.16), can reach values of the order of 100μ, the same order as the width of the domains themselves in materials such as iron.

Another recent advance has been in the use of materials with 'preferred orientation' of their crystal grains, chiefly the silicon steel used in large quantities in electrical power engineering. By suitable cold-rolling treatments it is possible to make a majority of grains crystallize with their axes close to some particular orientation, so that the whole material approximates (very crudely in many cases) to a single crystal. The most obvious desirable result of this is the raising of the 'knee' of the magnetization curve above the value it would have with random orientation of the grains, approximating to the [100] curve of fig. 2 rather than to some average of the three curves. It is also found, however, that the permeability of these materials is higher and their coercive force lower along the 'good' direction (the [100] axis in iron) than in other directions. The explanation of this effect is uncertain, because it involves questions of domain arrangement of the type left unanswered in § 4.5. It is possible, however, that the orienting effect of surfaces mentioned in § 6.13 may play the chief part.

---

* Recent work by Bozorth (1953) has resolved most of the difficulties. He shows that maximum permeability coincides with minimum magnetostriction in the easy directions.

Still further increases in permeability can be brought about in some materials by annealing them in a magnetic field (Bozorth and Dillinger, 1935; Goertz, 1951). By this means the magneto-strictive stresses associated with magnetization in one particular direction are relieved, so that, on cooling, the material will be more readily magnetizable in this direction (or antiparallel to it) than in other directions.

Materials for permanent magnets are required to have very different properties—high coercive force and high retentivity. These have usually been achieved by using materials with large internal stresses and many 'inclusions' of non-magnetic phases. The early magnet steels used carbon in the form of cementite—Fe_3C—to provide stress centres and inclusions. Improvements were later made by adding other elements, such as tungsten or cobalt, which prevented ageing and increased the coercive field somewhat, but most modern magnets are of 'dispersion-hardened' alloys, heat-treated so that separation into two phases is arrested just at the point when many tiny crystals have formed. These alloys, like the 'soft' magnetic materials, can be made anisotropic by annealing in a magnetic field, emphasizing the desirable properties in one direction at the expense of those in others.

One of the most striking recent developments is the application of Néel's theoretical results discussed in § 6.5 to the production of permanent-magnet materials. The theory indicates that high coercive forces should be obtained in any anisotropic ferro-magnetic, if only it is in the form of sufficiently small particles, each one consisting of a single domain. By using very finely divided iron a material has been produced with a coercive force up to 600 oersteds and a remanence of 5000 gauss. It is thus comparable with other permanent-magnet materials in its magnetic properties, while it is considerably easier to work mechanically.

144

CHAPTER VII

TIME EFFECTS

7.1. Introduction

Changes in magnetization are never instantaneous. Several different factors are known to contribute to the slowing up of magnetic changes, and the effects they produce cover a wide range of time, from many years to fractions of a microsecond. The methods of observing time effects may be divided into three classes: in the first, the field is changed abruptly and the subsequent changes in magnetization are studied by one means or another; in the second, an alternating magnetic field is used and the amplitude of the resulting magnetization is studied, particularly its dependence on frequency; the third method again employs an alternating field, but the quantity observed is either the phase of the magnetization or its equivalent, the energy lost by the field to the specimen.

The relations between the three methods can be illustrated by the simple example of a material in which the rate of change of magnetization is proportional to the difference between the actual magnetization and its equilibrium value, I_∞,

$$\frac{dI}{dt} = \frac{1}{\tau}(I - I_\infty). \tag{7.1}$$

We shall also suppose that the equilibrium magnetization is directly proportional to the applied field

$$I_\infty = \chi H.$$

If, then, we use the first method of investigation and change the field abruptly from o to H at time $t=$o, the magnetization will increase exponentially $\quad I = I_\infty(1 - e^{-t/\tau})$

$$= \chi H(1 - e^{-t/\tau}).$$

Observation of the changes in I will thus give the time constant τ directly. If we apply an alternating field

$$H = H_0 \sin \omega t, \tag{7.2}$$

the solution of equation (7.1) becomes

$$I = I_0 \sin(\omega t - \delta) = \frac{\chi H_0}{\sqrt{(1 + \omega^2 \tau^2)}} \sin(\omega t - \tan^{-1} \omega \tau). \quad (7.3)$$

The quantity observed in the second method is the 'effective susceptibility', χ_ω, which is given by

$$\chi_\omega = \frac{I_0}{H_0} = \frac{\chi}{\sqrt{(1 + \omega^2 \tau^2)}}. \quad (7.4)$$

In the third method the phase angle, δ, is measured, either directly or indirectly, through the energy lost by the field to the material in each cycle of magnetization. This energy, W, is given by

$$W = \int H dI = \frac{\omega \tau}{\sqrt{(1 + \omega^2 \tau^2)}} \chi H_0^2 = \chi H_0^2 \sin \delta. \quad (7.5)$$

It is, in fact, equal to the area of the elliptical loop obtained by plotting I against H from equations (7.2) and (7.3).

With this simple type of material it is clear that a single measurement by any one of the three methods will suffice to determine the constant τ and so define completely the time-dependent behaviour of the magnetization. Most materials, of course, do not behave in such a simple way; their properties can neither be described by a single time constant nor determined by a single experiment. Becker and Döring (1939, Ch. 19) have considered a more general case where different parts of the material possess different time constants, the volume with time constants between τ and $\tau + d\tau$ being defined by a function $g(\tau)$ which they assumed to be unity for a certain range of τ and zero outside that range. The calculated curves for the variation of χ_ω and W with frequency and for the change of magnetization after a sudden change in field agreed well with experiments on carbonyl iron, if suitable values of τ were used. In other cases the phenomena cannot be represented by any combination of 'relaxation' time constants, and other forms of time-dependence have to be introduced, such as resonance effects or the statistical effects of random impulses.

Snoek (1947) has pointed out that there are quite general mathematical relations between the three quantities that can be observed by the different experimental methods. If a complete experiment, i.e. one covering the whole range of frequencies or of time, is made by any one of the methods, the results of experiments by the other methods can be predicted. As a matter of experimental convenience it has been usual to use the first method at very low frequencies, the second at very high frequencies and the third for intermediate frequencies (the power-frequency range). This has sometimes led to the idea that three different effects are being observed; in fact, the effects are not independent and any mechanism that produces one must produce the other two as well.

Time variation of magnetic quantities other than the magnetization itself can be observed. It is well known, for example (Webb and Ford, 1934), that the reversible permeability of many materials changes rapidly immediately after they have received a magnetic or mechanical shock. These effects can be explained in the same sort of way as the rates of change of magnetization and will not be considered further.

The earliest observations on the changes of magnetization with time were those of Ewing (1885), who measured the magnetization in iron rods with a magnetometer, and observed effects, which he attributed to 'magnetic viscosity', lasting for up to 10 min. after a sudden change in field. These measurements were repeated and extended by Rayleigh (1887) and others, but with the advent of alternating-current engineering attention was diverted from the relatively long time effects, for which the slow change of magnetization could be observed directly, to effects of importance at power frequencies, for which the second or third method of observation had to be used. In power engineering the energy loss (W of equation (7.5)) is the quantity of most interest and the one usually measured. In communications engineering it is normally the variation of permeability with frequency that is of greatest importance. At very high frequencies the permeability decreases to unity, as indicated by the measurements of Rubens and Hagen (1903) on infra-red reflectivities, but before this happens, at high radio-frequencies, there are various resonance effects, which have been extensively studied recently.*

---

* Ferromagnetic resonance effects are not considered in any detail here as they have little direct bearing on domain structure. The subject has been reviewed recently by Kittel (1951).

7.2. Effects of eddy currents

The discussion of the various time effects just mentioned, and their interpretation in terms of mechanisms which cause slowness in magnetic changes, is complicated by the presence of a mechanism not in any way peculiar to ferromagnetics. This mechanism, ubiquitous, simple in principle but intricate in detail, is the eddy-current one. Every change in magnetization induces currents in the surrounding material, and the magnetic field of these currents will, by Lenz's law, oppose the change in magnetization. The calculation of the effects of these currents, if the material is assumed to be magnetically homogeneous, is a matter of ordinary electro-magnetic theory. The effects depend on the shape of the specimen and on the B-H relations for the material, simple results being obtained only when this relation is linear, i.e. when μ is constant. In this case Thomson (1892) obtained the following formulae for infinite flat sheets of thickness $2a$ placed in an alternating magnetic field $H = H_0 \cos \omega t$:

$$B = \mu H_0 \left(\frac{\cosh 2px + \cos 2px}{\cosh 2pa + \cos 2pa} \right)^{\frac{1}{2}} \cos(\omega t - \theta), \qquad (7.6)$$

with

$$\tan \theta = \frac{\sinh p(a-x)\sin p(a+x) + \sinh p(a+x)\sin p(a-x)}{\cosh p(a-x)\cos p(a+x) + \cosh p(a+x)\cos p(a-x)} \qquad (7.7)$$

and

$$p = 2\pi \sqrt{\frac{\mu\omega}{2\pi\rho}},$$

where B is the flux density at a distance x from the centre of the sheet, μ is the permeability of the material and ρ its resistivity. At high frequencies only the surface layers of the material, to depth of order $1/p$, are affected by the magnetic field; the interior is 'shielded' by the eddy currents circulating in the outer layers.

The effective permeability and energy loss can be calculated from equations (7.6) and (7.7) by integrating across the thickness of the sheet giving

$$\mu_\omega = \frac{\overline{B}}{H_0} = \frac{\mu}{\sqrt{2}\,pa} \left(\frac{\cosh 2pa - \cos 2pa}{\cosh 2pa + \cos 2pa} \right)^{\frac{1}{2}} \qquad (7.8)$$

and

$$W = \frac{\mu H_0^2}{8pa} \left(\frac{\sinh 2pa - \sin 2pa}{\cosh 2pa + \cos 2pa} \right). \qquad (7.9)$$

At low frequencies these results reduce to

$$\mu_\omega = \mu$$

and $$W = \mu^2 H^2 \pi \omega a^2 / 3\rho. \tag{7.10}$$

These have the same low-frequency form as equations (7.4) and
(7.5) with $\tau = 4\pi\mu a^2/3\rho$, so that to a first approximation the eddy
currents act as if they introduced a simple time lag into the relations between B and H.

If the appropriate values of μ, ρ and a are substituted in the
formulae, the resulting expressions for the frequency dependence
of permeability and loss agree well with the experimentally
observed dependence in many materials, showing that this normal
eddy-current mechanism is the chief factor determining rates of
change of magnetization. There are, however, many other materials
which do not exhibit this agreement, the observed losses or decrease of permeability with frequency being greater than those
calculated from the simple eddy-current theory. In power-
engineering practice such materials are said to exhibit an 'eddy-
current anomaly'; in the low-field region of communications
engineering the effect is called 'magnetic viscosity' or 'Nachwir-
kung'. Many attempts have been made to explain the discrepancy
by pointing out imperfections in the usual calculations of the
effects of eddy currents; the assumption that the B-H relation is
linear, for example, is obviously not fulfilled in many cases, and in
some materials the permeability may even vary systematically
through the thickness of a sheet. Allowance for these and other
complications in the eddy-current mechanism can be made by
various approximate methods (Brailsford, 1948b; Blake, 1949),
and may considerably modify the estimates of eddy-current effects
from equations (7.8) and (7.9), but when all allowance has been
made there remain many cases in which the calculated effects of
eddy currents are smaller than the time effects actually observed.
In such cases some other retarding mechanism besides the eddy-
current one must be operating.

The existence of time effects other than those caused by the
normal eddy currents can be shown still more clearly by studying
the effects as functions of suitable variables such as alternating flux

density, size and shape of specimen, and temperature. Thus Jordan (1924), followed, among others, by Ellwood (1935), investigated the dependence of the loss angle (δ of equation (7.5)) on flux density and frequency in high-permeability nickel-iron alloys. The loss angle due to eddy currents should vary linearly with frequency at low frequencies, and that due to hysteresis should vary linearly with the exciting field in the Rayleigh loop region, so that extrapolation to zero field and zero frequency should give zero loss angle. Jordan and Ellwood both found, however, that the extrapolated loss angle was very appreciably different from zero.

In electrical sheet steel as used for transformers the inability of normal eddy currents to explain all the time effects can be shown by experiments in which the thickness of the sheet is varied (Stewart, 1950). Equation (7.10) indicates that the eddy-current loss should be proportional to the square of the thickness. In fact, when the thickness is reduced the frequency-dependent loss decreases only slightly, showing that only a small part of it is caused by eddy currents.

An extensive study of time effects in carbonyl iron was made by Richter (1937), who measured the changes in magnetization after an abrupt change in field, and Schulze (1938), who measured the loss angle in alternating fields of 60–5000 c.s. Richter's results are shown in fig. 64. It is clear that such a strongly temperature-dependent effect cannot be due to eddy currents, since temperature could only affect the latter through the resistivity, ρ, which does not vary much with temperature. Schulze's results showed a similar temperature-dependence and could, in fact, be correlated with Richter's by assuming that a range of time constants was involved, as indicated on p. 145, the time constants depending approximately exponentially on the temperature, according to the law

$$\tau \approx 5 \times 10^{-15} \, e^{\theta/T} \text{ sec. with } \theta = 10,300°.$$

A large number of measurements on various materials have been made at higher frequencies; they have been reviewed by Strutt and Knol (1940) and by Allanson (1945). In nearly all materials it is found that the effective permeability falls off considerably for

frequencies of 1 mc./s. and over, indicating that the main magnetization processes are controlled by time constants of the order of
10^{-6} sec. At higher frequencies still the phenomenon of ferromagnetic resonance occurs, as mentioned above.

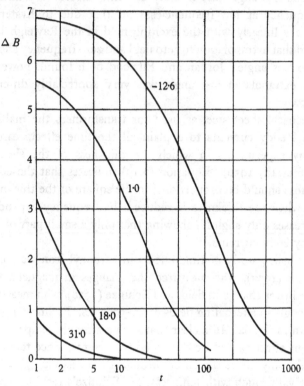

Fig. 64. Change of magnetization after a sudden change in field. The difference,
ΔB gauss, of the flux density from its equilibrium value is shown as a function
of time, t sec., for carbonyl iron at various temperatures (indicated alongside
the curves in ° C.). (After Richter, 1937.)

7.3. Mechanisms producing magnetic viscosity

Three different mechanisms producing time effects in magnetization have been put forward. It seems likely that combinations
of their effects can account for most of the observed phenomena.

The first effect is due to inertia of the elementary carriers of
magnetic moment, the electronic spins themselves. This naturally
leads to the resonance effects just mentioned, if the spins are

'elastically' constrained to some equilibrium position by the action of a magnetic or magnetocrystalline field. Calculations by Kittel (1951) and others show good agreement with experiment. The inertia of the spins may also mean that there is a definite limiting velocity for the movement of domains walls—which can be thought of as a wave motion of spin orientation—as has been suggested by Snoek (1947) and by Döring (1948 b). This effect probably contributes to the decrease of permeability at high radio-frequencies, but is unlikely to be important at frequencies below 1 mc./s. It has not been extensively studied as yet; it is only recently (Rado, Wright and Emerson, 1950) that it has been clearly distinguished from the second suggested mechanism, the action of eddy currents on a microscopic scale.

The eddy-current formulae, equations (7.8) and (7.9), quoted above were derived on the assumption that the magnetization varied smoothly and continuously throughout the material. This assumption is justified if we are considering averages over a region large enough to contain many domains, but it is clearly not valid for detailed calculations on the changes in the immediate neighbourhood of a domain wall. If we take account of the domain structure of ferromagnetics we may therefore expect eddy-current effects additional to the large-scale ones covered by formulae such as (7.8) and (7.9). Calculations for a realistic model of domain changes would be very complicated, but Becker (1938) has estimated the order of magnitude of the effect by treating simple models of isolated walls. If we consider a 180° wall, as in fig. 65, moving perpendicular to its plane so as to increase the volume of one domain and decrease that of the other, we can calculate the effect of the induced eddy currents as follows.

If the wall has area A and moves with velocity v, the combined magnetic moment of the two domains will change at a rate $2AI_s v$. If we consider an annulus of cross-section dr, $d\theta$ in the position (r, θ) the rate of change of flux through the annulus will be

$$\Phi = \frac{4AI_s v}{r^3} \int_0^\theta \cos \theta \, 2\pi r \sin \theta r \, d\theta$$

$$= \frac{8\pi AI_s v}{r} \sin^2 \theta,$$

and if the resistivity of the material is ρ, a current

$$\delta i = \frac{\Phi r\, d\theta\, dr}{2\pi r\rho \sin\theta} = \frac{4AI_s v}{r\rho} \sin\theta\, dr\, d\theta$$

will flow round the annulus, producing a field

$$\delta H = \frac{8\pi AI_s v}{\rho} \frac{\sin^2\theta}{r^2}\, dr\, d\theta$$

at the domain wall. By integrating over all values of θ and over values of r from a lower limit, R, to infinity, we obtain for the

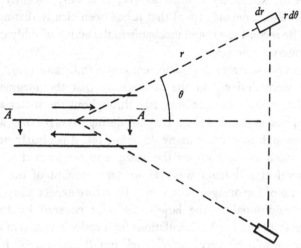

Fig. 65. Diagram for calculation of eddy currents induced by moving domain wall AA. Horizontal arrows show directions of magnetization, vertical arrows direction of wall movement.

total field produced by eddy currents acting at the domain wall

$$H_b = \frac{4\pi^2 AI_s}{\rho R} v. \tag{7.11}$$

As a rough approximation we may take R to be equal to the length of the domain, l, and also assume $A = l^2$ (this treatment is obviously incorrect unless the domains have all three dimensions about the same). The braking field, H_b, then becomes

$$H_b = \frac{4\pi^2 I_s}{\rho} lv$$

$$= bv, \quad \text{with} \quad \frac{4\pi^2 I_s}{\rho} l = b.$$

The equation of motion of a wall must therefore include the effects of this braking field, proportional to the wall velocity, as well as the effects of any applied field and of the obstacles to wall movement considered in the last chapter.

For reversible wall movements the effect of the obstacles is specified by the reversible susceptibility χ_r; if the material has $1/l$ walls, each of unit area, in a unit cube we can write

$$I = \chi_r H = 2I_s \frac{1}{l} x,$$

where x is the displacement of each wall and H is the sum of the applied field H_a and the braking field H_b. The equation of motion is thus

$$H_a - b \frac{dx}{dt} - \frac{2I_s}{\chi_r l} x = 0,$$

implying that all motion of the wall is damped, with a relaxation time constant

$$\tau_r = \frac{\chi_r l b}{2I_s} = \frac{2\pi^2 \chi_r l^2}{\rho}. \qquad (7.12)$$

The effects of such time constants on magnetization in alternating fields were discussed in § 7.1.

For irreversible wall movements the effect of the obstacles can be specified, in rather crude fashion, by two parameters, the excess field, H_0, that has to be applied to force a wall past the obstacles, and the distance, l, between successive obstacles. For the largest wall movements this distance can be identified with the length of a domain, l, already introduced, but it is now known (p. 87) that many smaller irreversible wall movements occur in which the distance of travel is short compared with the length of the domain. The effect of the eddy-current braking field on an irreversible movement will be to reduce the speed of the motion until $H_b = H_0$. We can thus write for the velocity of movement

$$V = \frac{H_0}{b},$$

and obtain an expression for the time, τ_i, taken to complete an irreversible movement,

$$\tau_i = \frac{l}{v} = \frac{l b}{H_0}$$

$$= \frac{4\pi^2 I_s l^2}{\rho H_0}. \qquad (7.13)$$

The precise effect of the eddy-current braking on the irreversible movements will clearly depend on the detailed form of the obstacles, but, in general terms, we can say that the effect will be negligible in applied fields whose frequency is much less than $1/\tau_i$, while for fields with frequencies much greater than $1/\tau_i$ the eddy currents will reduce the amplitude of the irreversible movements to a very small amount. As a rough approximation the excess field, H_0, in equation (7.13) may be taken as equal to the applied field; in general, irreversible wall movements will only contribute appreciably to changes in magnetization when the applied field is comparable with the coercive field. Comparison of equations (7.12) and (7.13) shows that in most ferromagnetic materials τ_r is considerably less than τ_i (when H_0 is taken as equal to the coercive field) and consequently irreversible wall movements are damped out at considerably lower frequencies than reversible ones.

Becker assumed the values $\chi_r = 10$, $\rho = 10^{-5}$ ohm cm., $l = 10^{-4}$ cm., $I_s = 1720$ gauss, $H_0 = 5$ oersteds, and so obtained time constants of order 10^{-9} sec. for reversible movements and 10^{-7} sec. for irreversible ones—the order of magnitude required to explain the observed radio-frequency effects (Allanson, 1945; Kreielsheimer, 1933). The values of H_0 and l may, however, vary widely from one material to another and give quite different values of τ_r and τ_i. In particular, in pure and strain-free iron or its alloys we may expect the domain size, l, to be large ($\sim 10^{-3}$ cm.) and the coercive force, H_0, small, so that τ_i becomes considerably larger ($\sim 10^{-4}$ sec.) and the small-scale eddy-current effects may account for a large part of the 'extra losses' and decrease in permeability observed at audio- and power-frequencies (Stewart, 1950). This increase of time effects as domain size is increased and coercive force reduced may set a limit to the reduction of losses in iron by purification, since the conditions required for low hysteresis loss appear to be the same as those that produce large 'extra' or viscosity loss.

The third possible cause of time effects in magnetization is the rearrangement of the atoms of the material. In some cases this may be a 'chemical' change, such as the slow precipitation of a new phase from solid solution. Such changes, which may take years to complete, are chiefly important for the irreversible changes in magnetic properties they produce, such as a decrease of permeability by ageing, rather than for any change in magnetization itself. Another type of atomic rearrangement is the gradual relief of stresses set up by mechanical or magnetic means.

Snoek (1939) has suggested that this mechanism may account for the time effects at low frequencies and low flux densities, particularly for the strongly temperature-dependent effects already mentioned. He points out that the wall between two domains is a region of considerable magnetostrictive stress, so that if the wall remains in one place for a time comparable with the mechanical relaxation time for stresses in the material, the material will be deformed in such a way as to relieve the magnetostrictive stresses and so create a minimum of potential energy for the wall at that point. Snoek pictures the wall as analogous to a ball rolling on a plastically deformed plane. If the ball remains in one place for a long time it will sink into the plane and then be more difficult to move quickly. A force too small to lift the ball from its hole will, however, still be able to move it slowly to one side or the other. This model explains both the frequency-dependence of permeability, and also the phenomenon of reversible decrease in permeability with time after a mechanical or magnetic shock.

The parameters required to characterize the effects are the time constant, τ, of the stress-relaxation process, and a measure of the magnitude of the relaxation, which may be expressed either as the field, H_r, required to move a wall after it has been in one place for time much longer than τ, or as the energy change, W_r, per unit volume, accompanying stress relaxation. The value of τ in any given material varies markedly with temperature, approximately according to an exponential law

$$\tau = \tau_0 e^{\theta/T}.$$

In some cases τ must be assumed to vary from place to place in the material. The values of H_r are normally a small fraction of an oersted (0·035 oersted for the 0·006 % carbon-steel used by Snoek), so that the magnetomechanical relaxation effects are only important for small changes of field. The magnetomechanical effect can clearly be related to the mechanical relaxation properties of the material, and Snoek has shown that changes in temperature or in composition affect both the magnetic and the mechanical time effects in iron in the same way.

Néel (1951) has recently suggested that the coupling between magnetic and atomic changes provided by Snoek's magnetostriction mechanism is not strong enough to explain all the observations on carbon-steel, but that the effect of migration of carbon atoms on the magnetocrystalline energy could act in a similar way and produce viscosities of the required magnitude.

A third way in which atomic motions can affect magnetization is through the irregular fluctuations of energy in their thermal agitation. These effects were first considered by Néel (1950) in connexion with the magnetization of small 'single-domain' particles. He showed that there was a finite, temperature-dependent probability that the magnetization of a particle in a given field less than its coercive field would reverse within a given time. The theory gives a very satisfactory account of the 'thermoremanent' behaviour of materials such as clays and bricks, which contain very small ferromagnetic particles. In solid ferromagnetics also, thermal fluctuations cause changes in magnetization, since they can sometimes allow a wall to pass an obstacle, although the steady applied field may not be enough to overcome the obstacle.

Néel has shown that these time effects caused by thermal fluctuations can be detected in a wide variety of materials and conditions, from the low-field experiments of Barbier (1950) on pure iron to those of Street and Woolley (1949) on Alnico in fields comparable with its coercive field. They can be distinguished from the other types of time effect by various characteristic differences of behaviour, notably their independence of temperature and of the frequency at which a loss angle is measured and the fact that the changes after successive changes of field do not obey the principle of superposition. The simplest way of representing them is by an equivalent field, $H(t)$; the change in magnetization at a time t after bringing the material to a new state is the product of this field and the irreversible part of the differential susceptibility of the material in its new state. Néel shows that $H(t)$ has approximately the form

$$H(t) = S(Q + \log t)$$

for large t.

This law fits the experiments well, the values of S ranging from 0·0005 oersted for pure iron to 1·7 oersteds for Alnico. The theoretical order of magnitude of S is given by setting

$$\frac{1}{8\pi} S^2 v = \tfrac{1}{2}kT,$$

where $\tfrac{1}{8}\pi S^2 v$ represents the energy of the field S in the volume, v, swept out by a domain wall moving from one obstacle to another (the volume of a Barkhausen jump), k is Boltzmann's constant and T the absolute temperature. If we put $v \approx 10^{-9}$ c.c. for soft iron and $v \approx 10^{-15}$ c.c. for permanent-magnet material, we get $S \approx 0.002$ oersted and $S \approx 2$ oersteds respectively, in agreement with the experimental results.

7.4. Experiments with simple domain arrangements

In the last section three mechanisms to explain time effects in magnetization were put forward. They all make use of the domain theory of ferromagnetic structure, and they all fail to give precise quantitative estimates of the effects because of the difficulties of specifying and allowing for the complexities of the domain arrangements. Each explanation has its range of application, the spin inertia effects at very high frequencies, the small-scale eddy currents at medium frequencies and the magnetomechanical interactions at low frequencies, but there is probably considerable overlapping, so that there may well be more than one cause of the observed time effects in any given case.

Comparison of theory and experiment would be easier if we had to deal with a simple and known arrangement of domains rather than the complex structures that usually occur. As already mentioned (§ 5.7), there are two experimental arrangements that produce simple domain structures, the stretched wires used by Sixtus and Tonks, and the picture-frame-shaped single crystals used by Williams.

In Sixtus and Tonks's experiment a wire with positive magnetostriction and low magnetocrystalline anisotropy is placed under tension, so that there are only two favourable directions of magnetization, in opposite directions along the wire. If the wire is originally all magnetized in one direction, the magnetization can be reversed by the passage along the wire of a 180° wall (as shown diagrammatically in fig. 43) by applying a uniform field H_0 to the whole wire and an extra local field, H_s, at one end to start the reversal. The speed of movement of the wall and its profile through the wire can be deduced by observation of the voltages induced in coils surrounding the wire. The profile found by Sixtus and Tonks is shown roughly in fig. 43, and typical results for the wall velocity are shown in fig. 66. The fact that the wall advances more quickly

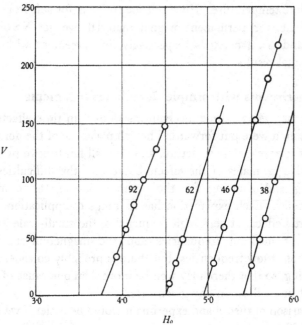

Fig. 66. Velocity of wall movement, V metres/sec., as a function of applied field, H_0 oersteds, in stretched nickel-iron wire, for various values of tension, indicated alongside the curves, in Kg./mm.² (After Sixtus, 1938.)

near the surface than at the centre of the wire suggests that eddy currents are providing a retarding field increasing towards the centre, and this idea is supported by the marked dependence of velocity on wire diameter, for large diameters.

In this case there is, of course, no distinction between 'large-scale' and 'small-scale' eddy currents; the eddy-current effect must be calculated from first principles for the particular circumstances. This can be done, provided the form of the profile of the advancing wall is known from theory or experiment. The experimental results agree with the assumption that the profile is determined by a balance between the surface energy of the boundary wall and the volume energy of the magnetization in the fields caused by the induced eddy currents, the free poles at the wall, and the external field. Exact calculations of the eddy-current effects are somewhat complicated (Sixtus and Tonks, 1932; Dijkstra and Snoek, 1949), but an approximate value for the wall velocity can be obtained by assuming a uniform distribution of eddy currents over the region of reversal, and equating the energy dissipated by these currents to that released by the reversal of the magnetization I_s in the external field H_0. The result is

$$V = \frac{k\rho}{8\pi^2 I_s a} H_0, \tag{7.14}$$

where ρ is the resistivity of the wire, a its radius, and k the length of wire occupied by the moving wall.

Equation (7.14) is in fair agreement with experiment for large diameters ($a > 0.05$ cm.) but gives too-large velocities in thin wires; this is equivalent to saying that when the eddy currents are large enough to keep the velocity below 20,000 cm./sec. they provide the main braking effect, but that when the eddy currents are small and would allow larger velocities, some other mechanism (suggested by Snoek (1947) to be a spin inertia one) restricts the velocity to about 20,000 cm./sec. Sixtus and Tonks's experiments have recently been repeated and extended by Dijkstra and Snoek (1949) and Ogawa (1949), and the presence of some frictional mechanism additional to the eddy-current one has been shown more clearly.

The picture-frame single-crystal method of obtaining simple domain structures has also been used to study the speed of wall movement (Stewart, 1951). A linear relation $V = A(H - H_0)$ between the velocity and the field was found, just as in Sixtus and Tonks's experiments, but the constant A was very much smaller— 6·0 cm./sec./gauss in a crystal of 3% silicon iron. Calculation showed, however, that this low speed could still be accounted for by the eddy-current mechanism because the dimensions of the crystal were much greater than the diameters in the stretched-wire experiments. The fact that A did not depend much on temperature indicated that mechanical relaxation effects were unimportant for the large wall movements that were studied. It seems likely that the relaxation effects could conveniently be investigated by similar experiments, using much smaller changes in magnetization.

The two methods of studying simple domain structures have thus yielded useful confirmation of the importance of eddy currents in controlling domain wall movement, but have not yet been made to give much information about the other factors that can limit the speed of movement. It is clear from the indirect evidence that such effects as spin inertia and mechanical relaxation can be important in many cases, but detailed calculation of these effects is not yet possible.

CHAPTER VIII

MAGNETIC AND THERMAL ENERGY CHANGES

8.1. Introduction

The various effects of inhomogeneities in ferromagnetics considered in Chapter VI complete the list of energy terms which have been found to be important in determining the equilibrium state of magnetization of a ferromagnetic. The list may have to be extended in the future, but at present it seems sufficiently comprehensive to explain the observed domain arrangements and changes of magnetization in nearly all cases. The chief inadequacies of present theory arise from the difficulty of expressing some of the energy contributions in terms of measurable quantities and of finding the configuration giving a minimum sum of all the different energy terms.

Throughout this book the expression 'energy' has been used somewhat loosely, referring to the quantity that must be minimized to obtain equilibrium, without specifying whether 'internal' or 'free' energy is meant. This is permissible for most purposes because, for temperatures well below the Curie point, temperature changes accompanying magnetization processes are small and the differences in the forms of 'energy' unimportant. Nevertheless, although the thermal effects of magnetization are small, they are of considerable practical and theoretical interest.

The earliest observations were of the irreversible heating which occurs when a ferromagnetic is taken round its hysteresis cycle. Although the rise in temperature for one cycle is small (of the order of 10^{-4}° C. for 'soft' material), it can be made easily measurable by performing many cycles. Reversible temperature effects, though they are of the same order of magnitude, are much harder to measure, since they cannot be amplified by repetition. It is only fairly recently that reliable experiments have been made (Bates and Weston, 1941; Bates and Harrison, 1948).

8.2. Irreversible heating effects: hysteresis

The connexion between the magnetization curve of a material and the heating produced by repeatedly reversing its magnetization was first stated by Warburg (1881), who showed that the thermal energy gained by the material for each cycle of magnetization is

$$W_H = \oint H \, dI, \tag{8.1}$$

i.e. the heat supplied (in ergs) is equal to the area of the hysteresis loop (in gauss × oersted units). This law can be deduced by con-

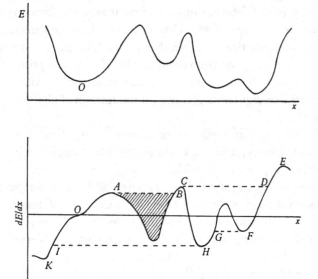

Fig. 67. Schematic representation of (a) energy and (b) gradient of energy as a function of wall displacement.

sidering the electrical energy that must be supplied to a solenoid containing the material in order to vary its magnetization cyclically, on the assumption that all the electrical energy 'lost' in the solenoid appears as heat in the magnetic material; it was verified approximately for iron by Warburg's experiments and, more exactly, by those of Bates and Weston (1941).

The hysteresis loss of a material is thus determined by its hysteresis loop, which, as we saw in Chapter VI, is normally dependent on local irregularities and could, in principle, be calculated if the irregularities were specified in sufficient detail. The relation between the obstacles (Chapter VI) caused by irregularities in the material and the hysteresis energy loss can be illustrated by reference to fig. 67, which represents schematically the variations in energy of a material as a domain wall moves in the x direction. In zero field the wall will be in equilibrium at an energy minimum such as O. If a field is applied favouring the direction of magnetization to the left of the wall rather than that to the right, the magnetic energy of the system will be decreased if the wall moves to the right, that is to say, a new energy term is added whose effect can be represented by raising the x-axis of fig. 67b by an amount proportional to the field, the new equilibrium position being fixed by the interesection of the dE/dx curve with this displaced axis. It can be seen that as the applied field is increased the equilibrium position moves steadily to the right as far as the maximum at A. During this process the work done by the magnetic field is stored in the material, the form in which it is stored depending on the mechanism which causes the obstacle A—it may be as increased surface energy of the wall or as any of the other energy forms considered in Chapter VI.

Beyond the maximum at A, the energy, for constant field, decreases as the wall moves to the right, so that A is a point of instability and beyond it a considerable movement of the wall will occur without increase of field. Further increase of field will move the wall on to C and thence, in another unstable jump, to D. Reduction of the field will not take the wall back along the same representative curve, but rather along the track $EDFGHIK\ldots$. The unstable movements A—B, C—D, etc., are thus irreversible. In each such movement the field supplies to the material energy in excess of that actually stored by the mechanism responsible for the curves of fig. 67. The excess energy is represented by the shaded area in fig. 67b for the jump from A to B. It is clear that the net amount of energy that must be supplied to move the wall from O to E and back to O again is represented by the area enclosed by the outer, broken curve of fig. 67b. Since distances along the x-axis, representing the position of the wall, are proportional to the change of magnetization and vertical distances of the equilibrium point are proportional to the field, fig. 67 b is in fact equivalent to the hysteresis loop of the material rotated by 90° from the position in which it is normally drawn.

8.3. Mechanisms for disposal of hysteresis energy

The hysteresis loss is thus seen to be the sum of the amounts of energy supplied in each 'jump' of the wall. Its magnitude is fixed by the curves of fig. 67, and the factors determining these curves have already been discussed, in Chapter VI. Fig. 67, however, defines only the amount of the energy lost and leaves unspecified the mechanism which disposes of it. During an unstable movement, such as that from A to B (fig. 67 b), the speed of movement of the wall would increase indefinitely if there were no forces, such as those considered in the preceding chapter, which restrict the speed. It is against these forces that the magnetic field does work during the unstable movement and, if the wall is to be brought to rest in a new equilibrium position, they must absorb the whole of the excess energy supplied by the field as the wall makes its jump.

Several different types of speed-controlling force were discussed in Chapter VII; for the 'jumps' occurring on the steep part of the magnetization curve in ordinary metallic ferromagnetics it is the small-scale eddy currents induced by the wall movement that play by far the most important part in controlling the speed of the wall and hence in dissipating the hysteresis energy.

As was mentioned in Chapter VII, the 'total loss' in a ferromagnetic subjected to a rapidly alternating field is greater than the hysteresis loss in a very slowly alternating one, and is usually resolved into three components, the hysteresis loss, the eddy-current loss and the 'viscosity' loss. If, as just stated, the hysteresis energy in ordinary ferromagnetic materials is dissipated by eddy currents, and if these eddy currents are also responsible for the viscosity of wall movement, some explanation is needed for the appearance of three experimentally distinguishable components of loss, each due to the eddy-current mechanism. The explanation, as given by Stewart (1951) and by Williams, Shockley and Kittel (1950), lies in the fact that the energy lost by eddy currents depends on the square of the current density. At low rates of change of magnetization we may picture each wall movement as occurring separately, its hysteresis energy being converted into heat by a system of local eddy currents occupying a limited region of space and time. If the magnetization is changed more rapidly, these regions will overlap, and hence the mean-square value of the eddy-current density will increase, with a consequent increase in energy lost. It is this increase in loss which appears as the 'eddy-current loss' in the classical calculation of the effects of eddy currents (equations (7.8)–(7.11)), taking no account of the domain structure of the material. The 'viscosity' component of loss appears when the applied field changes so rapidly that it increases appreciably during a single wall movement. It can be seen from fig. 67 that such an increase will increase the shaded area representing the energy lost during the wall movement and will therefore intensify the system of eddy currents associated with the movement, so that they dissipate the extra energy.

8.4. Reversibility of wall movements

If the scheme of fig. 67 is a good representation of hysteresis processes, then the loss of energy is due simply to the fact that the field is held constant during each jump and not adjusted to keep the wall in equilibrium at all times, that is to say, the dotted straight line between A and B is followed, rather than the full curve. If it were possible to reduce the field as soon as each maximum, such as A, was reached, and so to follow the equilibrium curve, the area of the hysteresis loop would shrink to zero and no energy would be lost in traversing it. To carry out this process would be extremely difficult in a normal ferromagnetic; the energy curve corresponding to a real domain wall probably has some hundreds of maxima rather than the three of fig. 67, and a normal specimen contains many thousands of domains, which cannot be controlled separately but must all be subjected to the same external field.

Single-crystal specimens can, however, be made (Williams and Shockley, 1949) in which the number of domains is drastically reduced, and with such specimens the idea of altering the field to follow the equilibrium curve of the domain walls can be partially realized (Stewart, 1951). The hysteresis loop obtained by an automatic method of keeping the field as close as possible to the equilibrium value is shown in fig. 68, as the full curve, indented where the wall is in unstable positions. The dotted line shows the hysteresis loop obtained without special control of the field. It was possible to follow the indentations of the hysteresis curve in this way because the wall movements in the specimens were very much slower than those in normal materials; a relatively crude feed-back

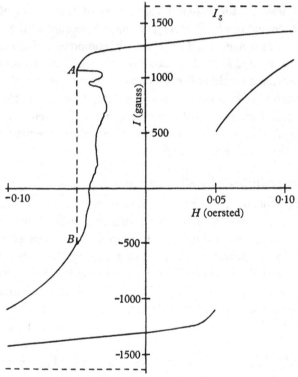

Fig. 68. *I-H* curve of a single-crystal specimen.

mechanism was thus able to adjust the field quickly enough to follow the curve shown. The slowness of the wall movement was a natural consequence of the large size of the domains, because the eddy-current control of speed is far stronger for large domains than for small.

It is clear that in the single-crystal specimen the hysteresis loss was appreciably reduced (by about 12%) by adjusting the field to follow the indented curve. If the field had been controlled by an apparatus with a quicker response, it is probable that a still more deeply indented curve would have been followed and that the loss would have been further reduced. It is unlikely, however, that the loss could have been reduced to zero, because the scheme of

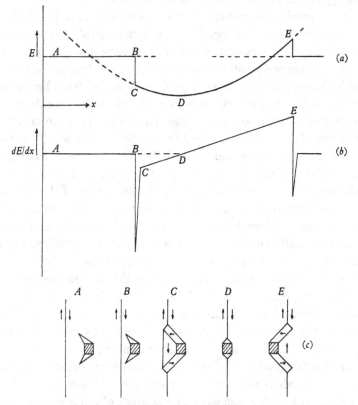

Fig. 69. Schematic representation of energy changes as a wall moves from left to right past a cubic non-magnetic inclusion.

fig. 67 is an over-simplification; for some types of obstacle to wall movement a single-valued energy curve, as shown in fig. 67, is an adequate representation, but for others there may be two or more states, with different energies, corresponding to the same position of the wall, so that the energy curve becomes double, as shown schematically in fig. 69. The sketches of domain arrangement shown in fig. 69 illustrate a type of obstacle with this behaviour, a cubic inclusion of non-magnetic material around which supplementary

closing domains are formed, as discussed in § 6.4. The arrangement of domains, and hence the energy of the system, depends on
the direction in which the wall is passing the inclusion. Obstacles
depending on magnetostrictive deformations (§ 6.6) can also give
energy curves of this type, if elastic hysteresis is present.

In materials whose obstacles have energy curves of the double-
valued type shown in fig. 69 we can distinguish two types of
hysteresis. One is associated with the movement of the main wall
in going from one position of equilibrium to another, that is to say,
with changes such as those from C to D in fig. 69. The energy
relationships in these changes have already been considered in
connexion with fig. 67, and it was shown that this component of
hysteresis loss could be reduced to zero by suitable adjustment of
the field acting on the domain wall. The second type of hysteresis
is associated with changes such as those from B to C in fig. 69,
where there is an irreversible change in the energy of the system,
not connected with any movement of the main wall but rather with
a change in the obstacle itself—in the example of fig. 69 the change
is in the small-scale domain structure round the inclusion, but it
could equally well be a mechanical change in the case of a stress
obstacle. The energy loss resulting from this second type of hysteresis is represented by the distance BC in fig. 69a and by the
area of the peak in the curve between B and C in fig. 69b. This
part of the hysteresis loss cannot be avoided by controlling the
magnetic field acting on the main wall and is, in this sense, the
more fundamental part of the loss. Apart from the case already
mentioned, where the 'wall-movement' part of the loss was found
to be at least 12 % of the whole, no estimate of the relative importance of the two components of the hysteresis has been made.

8.5. Reversible temperature changes

At temperatures near the Curie point the intrinsic magnetization
changes rapidly with temperature and changes of magnetization
can produce large thermal effects. This magneto-caloric effect
will not be discussed here, as it has little connexion with the domain
structure of ferromagnetics. Below the Curie point the temperature changes caused by magnetization are small and it is a matter
of considerable difficulty to measure them accurately. Many of the

earlier experiments gave misleading results, but recent work, particularly that of Bates and his collaborators, has given a large quantity of reliable data. Typical results for a specimen of iron are shown in fig. 70. The irreversible heating due to hysteresis is represented by the height of the final point above the starting-point. It can be shown that nearly all this irreversible heating occurs in the range of field from − 20 to + 20 oersteds, as would be

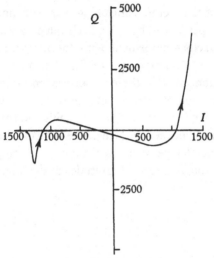

Fig. 70. Heat energy evolved, Q ergs/c.c., as a function of magnetization, I gauss, in traversing half of a hysteresis cycle for annealed Armco iron. (After Bates and Harrison, 1948.)

expected from the hysteresis loop. Curves of the temperature changes in other materials show considerable diversity, but recent work by Stoner and Rhodes (1949) gives a satisfactory thermo-dynamic explanation of the results in all cases where sufficient data are available.

The thermodynamics of ferromagnetics has been treated by many authors, notably Guggenheim (1936), Livens (1947) and Stoner (1935, 1937), for a variety of systems with a wide range of notations, definitions and approximations. For the present problem, however, Stoner obtains the simple basic equation

$$\left(\frac{\partial T}{\partial H}\right)_s = -\frac{T}{\rho C_H}\left(\frac{\partial I}{\partial T}\right)_H \qquad (8.2)$$

for the adiabatic change of temperature with field (ρ = density, C_H = specific heat at constant field). If the change of magnetization with temperature (in constant field) were known, then the change of temperature with field could be predicted. So far no suitable measurements of $(\partial I/\partial T)_H$ have been published. (Equation (8.2) applies only to reversible changes, so that care must be taken to include only the reversible part of $(\partial I/\partial T)_H$ in any calculation.) It is, however, possible to deduce values of $(\partial I/\partial T)_H$ from a knowledge of the factors controlling magnetization and of their temperature-dependence; by using such values Stoner and Rhodes obtained satisfactory agreement with the measured temperature changes. In the calculation of $(\partial I/\partial T)_H$ they allowed for changes of magnetization by the three processes envisaged in domain theory, the variation of intrinsic magnetization, I_s, the rotation of the intrinsic magnetization in whole domains and the displacement of the boundaries between domains. The contributions of the three processes to the change of magnetization with temperature were found to be of roughly equal magnitudes in the metals considered.

REFERENCES

AKULOV, N. S. (1929). *Z. Phys.* **57**, 249.
AKULOV, N. S. (1931 a). *Z. Phys.* **69**, 78.
AKULOV, N. S. (1931 b). *Z. Phys.* **69**, 822.
ALLANSON, J. T. (1945). *J. Instn Elect. Engrs*, **92** (iii), 247.
BARBIER, J. C. (1950). *C.R. Acad. Sci., Paris*, **230**, 1040.
BATES, L. F. and HARRISON, E. G. (1948). *Proc. Phys. Soc.* **60**, 213, 225.
BATES, L. F. and MEE, C. D. (1952). *Proc. Phys. Soc.* A, **64**, 129, 140.
BATES, L. F. and NEALE, F. E. (1950). *Proc. Phys. Soc.* A, **63**, 374.
BATES, L. F. and WESTON, J. C. (1941). *Proc. Phys. Soc.* A, **55**, 188.
BATES, L. F. and WILSON, G. W. (1951). *Proc. Phys. Soc.* A, **64**, 691.
BATES, L. F. and WILSON, G. W. (1953). *Proc. Phys. Soc.* A, **66**, 819.
BECK, K. (1918). *Vjschr. naturf. Ges. Zurich*, Jahrg. 63.
BECKER, R. (1930). *Z. Phys.* **62**, 253.
BECKER, R. (1932). *Phys. Z.* **33**, 905.
BECKER, R. (1938). *Phys. Z.* **39**, 856.
BECKER, R. and DÖRING, W. (1939). *Ferromagnetismus*. Berlin: Springer.
BECKER, R. and KERSTEN, M. (1930). *Z. Phys.* **64**, 660.
BICKFORD, L. R. (1950). *Phys. Rev.* **78**, 449.
BIDWELL, S. (1890). *Proc. Roy. Soc.* **47**, 469.
BITTER, F. (1931). *Phys. Rev.* **38**, 1903.
BLAKE, L. R. (1949). *J. Instn Elect. Engrs*, **96** (ii), 705.
BLOCH, F. (1932). *Z. Phys.* **74**, 295.
BOZORTH, R. M. (1937). *J. Appl. Phys.* **8**, 575.
BOZORTH, R. M. (1951). *Ferromagnetism*. New York: van Nostrand.
BOZORTH, R. M. (1953). *Rev. Mod. Phys.* **25**, 42.
BOZORTH, R. M. and DILLINGER, J. (1935). *Physics*, **6**, 279.
BOZORTH, R. M., MASON, W. P., MCSKIMMIN, H. J. and WALKER, J. G. (1949). *Phys. Rev.* **75**, 1954.
BOZORTH, R. M. and WALKER, J. G. (1950). *Phys. Rev.* **79**, 888.
BOZORTH, R. M. and WILLIAMS, H. J. (1945). *Rev. Mod. Phys.* **17**, 72.
BRAILSFORD, F. (1948 a). *Magnetic Materials*. London: Methuen.
BRAILSFORD, F. (1948 b). *J. Instn Elect. Engrs*, **95** (ii), 38.
BROOKS, H. (1940). *Phys. Rev.* **58**, 909.
BROWN, W. F. (1941). *Phys. Rev.* **60**, 139.
BRUKHATOV, N. L. and KIRENSKY, N. V. (1937). *Phys. Z. Sowjet*, **12**, 602.
CZERLINSKI, E. (1932). *Ann. Phys., Lpz.* (v), **13**, 80.
DIJKSTRA, L. J. and SNOEK, J. L. (1949). *Phillips Res. Rep.* **4**, 334.
DÖRING, W. (1936). *Z. Phys.* **103**, 560.
DÖRING, W. (1938). *Z. Phys.* **108**, 137.
DÖRING, W. (1948 a). *Z. Phys.* **124**, 501.
DÖRING, W. (1948 b). *Z. Naturforsch.* **3** (a), 373.
DÖRING, W. and HAAKE, H. (1938). *Phys. Z.* **39**, 865.
ELLWOOD, W. B. (1935). *Physics*, **6**, 295.
ELMORE, W. C. (1938). *Phys. Rev.* **54**, 1092.

EWING, J. A. (1885). *Phil. Trans.* **176**, 554.

FÖRSTER, F. and STAMBKE, K. (1941). *Z. Metallk.* **33**, 97.

FOWLER, C. A. and FRYER, E. M. (1952). *Phys. Rev.* **86**, 426.

GANS, R. (1932). *Ann. Phys., Lpz.* (v), **15**, 28.

GANS, R. and CZERLINSKI, E. (1932). *Schr. Königsb. gelehrt. Ges. Naturw. Kl.* **9**, 1.

GERMER, L. H. (1942). *Phys. Rev.* **62**, 295.

GOENS, E. and SCHMID, E. (1931). *Naturwissenschaften*, **19**, 520.

GOERTZ, M. (1951). *J. Appl. Phys.* **22**, 964.

GORTER, C. J. (1933). *Nature, Lond.*, **132**, 517.

GRABBE, E. M. (1940). *Phys. Rev.* **57**, 728.

GRIFFITHS, J. H. E. (1946). *Nature, Lond.*, **158**, 670.

GUGGENHEIM, E. A. (1936). *Proc. Roy. Soc.* A, **135**, 49, 70.

VON HAMOS, L. and THIESSEN, P. A. (1931). *Z. Phys.* **71**, 442.

HEISENBERG, W. (1928). *Z. Phys.* **49**, 619.

HEISENBERG, W. (1931). *Z. Phys.* **69**, 287.

HOLSTEIN, T. and PRIMAKOFF, H. (1940). *Phys. Rev.* **58**, 1098.

HOLSTEIN, T. and PRIMAKOFF, H. (1941). *Phys. Rev.* **59**, 388.

HONDA, K. and KAYA, S. (1926). *Sci. Rep. Tôhoka Univ.* **15**, 721.

HONDA, K. and MASIYAMA, Y. (1926). *Sci. Rep. Tôhoku Univ.* **15**, 755.

HONDA, K., MASUMOTO, H. and KAYA, S. (1928). *Sci. Rep. Tôhoku Univ.* **17**, 118.

HONDA, K. and SHIMIZU, S. (1902). *Phys. Z.* **3**, 375.

HOZELITZ, K. (1952). *Ferromagnetic properties of metals and alloys.* Oxford: University Press.

JORDAN, J. P. (1924). *Elekt. NachrTech.* **1**, 7.

JOULE, J. P. (1847). *Phil. Mag.* **30**, 76, 225.

KAYA, S. (1928). *Sci. Rep. Tôhoku Univ.* **17**, 111, 639, 1157.

KAYA, S. (1933). *Z. Phys.* **84**, 705.

KAYA, S. (1934). *Z. Phys.* **89**, 796; **90**, 551.

KAYA, S. and TAKAKI, H. (1936). *Honda Anniv. Vol.* p. 314.

KERSTEN, M. (1931a). *Z. Phys.* **71**, 553.

KERSTEN, M. (1931b). *Z. Tech. Phys.* **12**, 665.

KERSTEN, M. (1933). *Z. Phys.* **82**, 723.

KERSTEN, M. (1938). *Probleme der Technische Magnetisierungskurve.* Berlin: Springer.

KERSTEN, M. (1943). *Grundlagen einer Theorie der ferromagnetischen Hysterese und der Koerzitivkraft.* Leipzig: Hirzel.

KIMURA, R. and OHNO, K. (1934). *Sci. Rep. Tôhoku Univ.* **23**, 359.

KIRCHNER, H. (1936). *Ann. Phys., Lpz.* (v), **27**, 49.

KITTEL, C. (1946). *Phys. Rev.* **70**, 965.

KITTEL, C. (1949a). *Phys. Rev.* **76**, 1527.

KITTEL, C. (1949b). *Rev. Mod. Phys.* **21**, 541.

KITTEL, C. (1951). *J. Phys. Radium*, **12**, 292.

KONDORSKY, E. (1937). *Phys. Z. Sowjet.* **11**, 597.

KONDORSKY, E. (1938). *Phys. Rev.* **53**, 1022.

KONDORSKY, E. (1942). *J. Phys. U.S.S.R.* **6**, 93.

KORNETSKI, M. (1933). *Z. Phys.* **87**, 560.

KREIELSHEIMER, K. (1933). *Ann Phys., Lpz.* (v), **17**, 293.

LANDAU, L. and LIFSHITZ, E. (1935). *Phys. Z. Sowjet.* **8**, 153.

LAWTON, H. (1949*a*). *Proc. Camb. Phil. Soc.* **45**, 145.

LAWTON, H. (1949*b*). Thesis, Camb. Univ.

LAWTON, H. and STEWART, K. H. (1948). *Proc. Roy. Soc.* A, **193**, 72.

LAWTON, H. and STEWART, K. H. (1950). *Proc. Phys. Soc.* A, **63**, 848.

LEE, E. W. (1953). *Proc. Phys. Soc.* A, **66**, 623.

LIFSHITZ, E. (1944). *J. Phys. U.S.S.R.* **8**, 337.

LILLEY, B. A. (1950). *Phil. Mag.* (VII), **41**, 792.

LIVENS, G. H. (1947). *Phil. Mag.* (VII), **38**, 453.

LOVE, A. E. H. (1926). *A Treatise on the Mathematical Theory of Elasticity.* Cambridge University Press.

McKEEHAN, L. W. (1937). *Phys. Rev.* **51**, 136.

MAHAJANI, G. S. (1929). *Phil. Trans.* A, **228**, 63.

MARTON, L. (1949). *J. Appl. Phys.* **20**, 1258.

MASIYAMA, Y. (1928). *Sci. Rep. Tôhoku Univ.* **17**, 945.

MEE, C. D. (1950). *Proc. Phys. Soc.* A, **63**, 922.

NÉEL, L. (1942). *Cah. Phys.* **12**, 1; **13**, 1.

NÉEL, L. (1944*a*). *J. Phys. Radium*, **5**, 241.

NÉEL, L. (1944*b*). *Cah. Phys.* **25**, 1.

NÉEL, L. (1946). *Ann. Univ. Grenoble*, **22**, 299.

NÉEL, L. (1947). *C.R. Acad. Sci.*, Paris, **224**, 1488, 1550.

NÉEL, L. (1948*a*). *J. Phys. Radium* (VIII), **9**, 184.

NÉEL, L. (1948*b*). *J. Phys. Radium* (VIII), **9**, 193.

NÉEL, L. (1948*c*). *Ann. Phys.*, Paris, **3**, 137.

NÉEL, L. (1950). *J. Phys. Radium*, **11**, 49.

NÉEL, L. (1951). *J. Phys. Radium*, **12**, 334.

NEWTON, R. R. and KITTEL, C. (1948). *Phys. Rev.* **74**, 1601.

OGAWA, S. (1949). *Sci. Rep. Tôhoku Univ.* **1**, 53.

OSBORN, J. A. (1945). *Phys. Rev.* **67**, 351.

POLLEY, H. (1939). *Ann. Phys.*, Lpz. (v), **36**, 625.

PREISACH, F. (1932). *Phys. Z.* **33**, 913.

PREISACH, F. (1935). *Z. Phys.* **94**, 277.

RADO, G. T., WRIGHT, R. W. and EMERSON, W. M. (1950). *Phys. Rev.* **80**, 273.

RAYLEIGH, LORD (1887). *Phil. Mag.* **23**, 225.

RICHARDS, C. E., BUCKLEY, S. E., BARDELL, P. R. and LYNCH, A. C. (1950). *J. Instn Elect. Engrs*, **97** (ii), 236.

RICHTER, G. (1937). *Ann. Phys.*, Lpz. (v), **29**, 605.

RUBENS, H. and HAGEN, E. (1903). *Ann. Phys.*, Lpz. (IV), **11**, 873.

SCHARFF, G. (1935). *Z. Phys.* **97**, 73.

SCHULZE, H. (1938). *Probleme der Technische Magnetisierungskurve.* Berlin: Springer.

SHOENBERG, D. and WILSON, A. J. C. (1946). *Nature, Lond.*, **157**, 548.

SIXTUS, K. J. (1935). *Phys. Rev.* **48**, 425.

SIXTUS, K. J. (1937). *Phys. Rev.* **51**, 870.

SIXTUS, K. J. (1938). *Probleme der technische Magnetisierungskurve*, ed. Becker. Berlin: Springer.

SIXTUS, K. J. and TONKS, L. (1931). *Phys. Rev.* **37**, 930.

SIXTUS, K. J. and TONKS, L. (1932). *Phys. Rev.* **42**, 419.

SIXTUS, K. J. and TONKS, L. (1933). *Phys. Rev.* **43**, 70, 931.

SIZOO, G. J. (1929). *Z. Phys.* **56**, 649.

SNOEK, J. L. (1939). *Physics*, **6**, 161, 321, 591, 797.

SNOEK, J. L. (1947). *New Developments in Ferromagnetic Materials.* Amsterdam: Elsevier.

STEWART, K. H. (1949). *Proc. Camb. Phil. Soc.* **45**, 296.

STEWART, K. H. (1950). *J. Instn Elect. Engrs*, **97** (ii), 121.

STEWART, K. H. (1951). *J. Phys. Radium*, **12**, 325

STONER, E. C. (1935). *Phil. Mag.* (VII), **19**, 565.

STONER, E. C. (1937). *Phil. Mag.* (VII), **23**, 833.

STONER, E. C. (1945). *Phil. Mag.* (VII), **36**, 803.

STONER, E. C. (1948). *Rep. Progr. Phys.* **11**, 43.

STONER, E. C. (1950). *Rep. Progr. Phys.* **13**, 83.

STONER, E. C. and RHODES, P. (1949). *Phil. Mag.* (VII), **40**, 481.

STONER, E. C. and WOHLFARTH, E. P. (1948). *Phil. Trans.* A, **240**, 599.

STREET, R. and WOOLLEY, J. C. (1949). *Proc. Phys. Soc.* A, **62**, 562.

STRUTT, M. J. O. and KNOL, K. S. (1940). *Physica*, **7**, 635.

SUCKSMITH, W., POTTER, H. H. and BROADWAY, L. (1928). *Proc. Roy. Soc.* A, **117**, 476.

TAKAGI, M. (1939). *Sci. Rep. Tôhoku Univ.* **28**, 20.

TEBBLE, R. S. and NEWHOUSE, V. L. (1953). *Proc. Phys. Soc.* A, **66**, 633.

TEBBLE, R. S., SKIDMORE, I. C. and CORNER, W. D. (1950). *Proc. Phys. Soc.* A, **63**, 739.

THOMSON, J. J. (1892). *Electrician*, **28**, 559.

VAN VLECK (1945). *Rev. Mod. Phys.* **17**, 27.

VAN VLECK (1947). *Ann. Inst. Poincaré*, **10**, 57.

VON ENGEL, A. and WILLS, M. S. (1947). *Proc. Roy. Soc.* A, **188**, 464.

WALKER, J. G., WILLIAMS, H. J. and BOZORTH, R. M. (1949). *Rev. Sci. Instrum.* **20**, 947.

WARBURG, E. (1881). *Ann. Phys., Lpz.* (III), **13**, 141.

WEBB, C. E. and FORD, L. H. (1934). *J. Instn Elect. Engrs*, **75**, 787.

WEBSTER, W. L. (1925). *Proc. Roy. Soc.* A, **109**, 570.

WEBSTER, W. L. (1930). *Proc. Phys. Soc.* **42**, 431.

WEISS, P. (1905). *J. Phys. Radium*, **4**, 469, 829.

WEISS, P. (1907). *J. Phys. Radium*, **6**, 661.

WEISS, P. (1910). *J. Phys. Radium*, **9**, 373.

WEISS, P. and FORRER, R. (1926). *Ann. Phys., Paris* (x), **5**, 153.

WEISS, P. and FORRER, R. (1929). *Ann. Phys., Paris* (x), **12**, 279, 316.

WILLIAMS, H. J. (1937a). *Rev. Sci. Instrum.* **8**, 56.

WILLIAMS, H. J. (1937b). *Phys. Rev.* **52**, 747, 1004.

WILLIAMS, H. J., BOZORTH, R. M. and SHOCKLEY, W. (1949). *Phys. Rev.* **75**, 155.

WILLIAMS, H. J., FOSTER, F. G. and WOOD, E. A. (1951). *Phys. Rev.* **82**, 119.

WILLIAMS, H. J. and SHOCKLEY, W. (1949). *Phys. Rev.* **75**, 178.

WILLIAMS, H. J., SHOCKLEY, W. and KITTEL, C. (1950). *Phys. Rev.* **80**, 1090.

WILLIAMS, H. J. and WALKER, J. G. (1951). *Phys. Rev.* **83**, 634.

YAMAMOTO, M. and IWATA, T. (1951). *Phys. Rev.* **81**, 887.

INDEX

Alternating fields, 144 ff.
Anisotropy constant
  measurement, 36 ff.
  values, 9, 40
Approach to saturation, 34 ff.

Barkhausen effect, 92
'Bitter patterns', 81 ff.

Closure, domains of, 70 ff.
  energy in, 76 ff.
Coercive force, 116, 121, 126, 129,
  132
Curie point, temperature, 3

Distortion of crystals, 42

'Easy directions', 6, 18, 48, 74
Eddy currents, 147 ff., 158, 164
Elastic properties, effect of magneto-
  striction on, 64 ff.

Ferrimagnetism, 5
Field, demagnetizing, effects on
  cylindrical rods, 12, 30
  disk specimens, 17, 25 ff.
  domain boundaries, 67 ff., 105, 114,
    119, 131
  external field, 7
  free energy, 68, 73
  grains in polycrystals, 32
  iron under tension, 60
  magnetostriction, 53
  spheres, 10
Field, internal (or molecular), 3, 4

Hysteresis, 161 ff.
  in low fields, 137 ff.

Imperfections, effects of, 14, 108 ff.
Inclusions, 112 ff., 167
  distribution, 131
Internal stresses, see Stresses

'Knee' of magnetization curve, 16, 31,
  142

Loss in alternating fields, 145, 147 ff.,
  154, 164

Magnetization,
  micro- or intrinsic, spontaneous,
    2 ff., 23, 35, 42, 170
  macro-, 2
  rate of change, 144 ff.
  rotation, 9, 13, 18, 34, 55, 118, 124,
    127, 170
  thermal effects, 161, 168
  work of, 18, 38, 125, 129
Magnetization curves
  for polycrystals, 33
  for single crystals, 8, 18 ff.
  for stressed materials, 55 ff.
  in low fields, 137 ff.
  near saturation, 34
Magnetocrystalline anisotropy, 8,
  17 ff., 59, 73, 76, 95, 118, 140, 156
  coefficients, 19
  measurement of constants, 36 ff.
  physical origin, 24
Magnetostriction, 9, 41 ff., 73, 81, 94,
  101, 110, 119, 124, 128, 140, 155
  coefficients, 46
  dependence on magnetization, 48 ff.
  effect on elasticity, 64 ff.
  effect of tension, 58
  physical interpretation, 66

Patterns, domain
  'fir tree', 85
  in '[110] rod', 74 ff.
  in low fields, 80
  in uniaxial crystal, 71
  mechanism of formation, 89 ff.
  near inclusions, 116, 167
  observation, 81 ff., 92
  on cobalt, 92
  on iron, 83 ff.
  on nickel, 91
Permanent magnets, 143
Permeability, 15, 140
  at high frequencies, 149
  initial, 121, 128, 138
  time change, 146
Plastic deformation, 17, 82, 128
Polishing of specimens, 82, 83
Polycrystals
  magnetization curves, 32 ff.
  work of magnetization, 38

Remanence, 125
Resonance, ferromagnetic, 40, 146

Shape, effect of, 17, 53, 119
Single crystals
  experiments, 17 ff., 49 ff., 83, 86, 103, 157, 165
  initial susceptibility, 135
  magnetization curves, 18, 25
  magnetostriction, 48 ff.
  production, 17
  torque, 39, 136
Single-domain particles, 118, 156
Stress anisotropy, 9
Stress
  distortion of crystal by, 42
  effect on magnetic properties, 53 ff.
  effect on wall energy, 101, 110
Stress, internal, 22, 34, 64, 65, 81, 155
  effect on domain arrangements, 109 ff.
  effect on permeability, 127, 129
  lowest value, 128
Susceptibility, initial, 56, 121, 128
  anisotropy of, 134

Temperature changes during magnetization, 168

Tension, effect of
  on iron, 60 ff.
  on magnetostriction, 58
  on nickel, 54 ff.
  on Permalloy, 57
Thermoremanence, 156
Time constants for magnetic change, 144 ff.
Torque on crystals, 39, 136

Villari reversal, 51
Viscosity, magnetic, 150 ff., 164
Volume, variation of, with magnetization, 44, 52

Walls, domain, 6, 93 ff.
  deformation, 131
  effect on domain spacing, 72
  energy of, 99, 103
  hindrances to movement, 13, 108 ff.
  measurement of surface energy, 103 ff.
  movement, 151, 165, 170
  orientation, 67 ff.
  velocity of movement, 151 ff., 158
  width, 100

Printed in the United States
By Bookmasters